THE CUT / READING BATAILLE'S *HISTOIRE DE L'ŒIL*

THE CUT /
READING BATAILLE'S
HISTOIRE DE L'ŒIL

Patrick ffrench

*A British Academy
Postdoctoral Fellowship Monograph*

Published for THE BRITISH ACADEMY
by OXFORD UNIVERSITY PRESS

Oxford University Press, Great Clarendon Street, Oxford OX2 6DP

Oxford New York
Athens Auckland Bangkok Bogota Bombay
Buenos Aires Calcutta Cape Town Dar es Salaam
Delhi Florence Hong Kong Istanbul Karachi
Kuala Lumpur Madras Madrid Melbourne
Mexico City Nairobi Paris Singapore
Taipei Tokyo Toronto Warsaw

and associated companies in
Berlin Ibadan

Published in the United States by
Oxford University Press Inc., New York

British Library Cataloguing in Publication Data
Data available

ISBN 0-19-726200-7

Typeset by Intype London Ltd
Printed in Great Britain
on acid-free paper by
Creative Print and Design (Wales) Ebbw Vale

Contents

Acknowledgements	vii
Georges Bataille	ix
1. / the cut	1
2. / obtuse	4
3. / structural(ist)	6
4. / œ I	14
5. / *informe*	17
6. / symptom	27
7. / spilled ink	33
8. / the influence of anxiety	39
9. / the institution	42
10. / centrifugal / centripetal	52
11. / a map	57
12. / *auch*	79
13. / framing	82
14. / narration	85
15. / roundness–liquidity–light	94
16. / punning	98
17. / œ II	101
18. / eye as figure	101
19. / *rose et noire*	103

20. / the apparatus   114

21. / out of this world!   125

22. / *corrida!*   132

23. / don juan   148

24. / cure   157

25. / writing blind / reading blind   175

Bibliography   177

Index   179

Acknowledgements

I am grateful to the British Academy for making the publication of this book possible, through their Postdoctoral Monograph series, and for the Postdoctoral Fellowship I held from 1993 to 1996. The Department of French at University College London has offered a productive context for the research which led to this book. Professor Timothy Mathews has offered useful advice on its composition. I am, as always, indebted to Professor Annette Lavers for her inspirational example, and to Roland-François Lack, Burhan Tufail, Bruno Sibona and Anita Phillips for their critical voices.

Georges Bataille

Georges Bataille (1897–1963) is a writer whose importance for the culture of modernity is increasingly recognized. Barely known outside a restricted group of associates and readers until perhaps the 1970s, when his *Œuvres complètes* began to be published and many of his key texts were translated,[1] Bataille had nevertheless played a crucial role in many of the major debates within French cultural life, from the 1920s— in relation to Surrealism—to the post-war period, in polemical debate with Sartre.[2] The structuralist and post-structuralist epoch of the 1960s onwards recognizes Bataille as a formative influence, although this has only recently been accounted for in the reception of French theory in the Anglophone con-

---

[1] The first volume of the *Œuvres complètes* was published in 1970, edited by Denis Hollier and introduced by Michel Foucault. So far twelve volumes have appeared. The obituary issue of *Critique*, the review Bataille founded in 1946, gives an extensive bibliography of articles, among them important texts by Blanchot, Klossowski, Duras and Sartre, but no book-length study had appeared by then. The review *La Ciguë* had published the only special issue on Bataille in 1958, featuring texts by Duras, Leiris, Fautrier, René Char, Malraux, Masson and Louis René des Forêts. Philippe Sollers, principal animator of the review *Tel Quel* which was to play a large part in the reactivation of Bataille's influence in the 1960s, recalls in an interview with the author that up until 1970 it was extremely difficult to come by copies of Bataille's works, and that Bataille remained a relatively unknown figure. As concerns the English translations of Bataille, Neugroschel's version of *Histoire de l'œil* first appeared in 1977 (New York: Urizen Books), although a previous translation by Austryn Wainhouse for Olympia Press, now unavailable, had appeared in 1958. Harry Mathews's translation of *Le bleu du ciel* first appeared in 1978 (New York: Urizen Books). Translations of the major discursive writings were only to appear in the mid- to late '80s.

[2] Sartre's critical article on Bataille, 'Un nouveau mystique', first appeared in the *Cahiers du Sud* in 1943 and was reprinted in *Situations I* (Paris: Gallimard, 1947). Bataille would respond to Sartre in his *Sur Nietzsche* of 1945. Bataille's *La littérature et le mal* would include a number of essays in which he explicitly responds to Sartre's 'biographies' of Baudelaire and Genet.

text.[3] This relative neglect contrasts with the present situation, in which Bataille's mark upon a number of different discourses, from anthropology to art history to philosophy and politics, is being investigated with unbounded enthusiasm at the discovery of a voice which undermines, which ruptures, which inscribes a cut in the corpus of knowledge.

Beyond, or perhaps below, this once muted but now jubilant influence, is a sense of anxiety, however. Bataille was also the writer of fictional texts, which have not ceased to haunt writers and artists alike.[4] Texts such as *Madame Edwarda*, *Le bleu du ciel*, *L'abbé C*, *Ma mère*, and the first—*Histoire de l'œil*, written in 1927 and published in 1928 under the pseudonym Lord Auch—have exerted a hidden but irreducible influence on the literary and artistic imagination of the century. I use the term *anxiety* here to indicate the profound sense of unease which a text like *Histoire de l'œil* induces in its reader, and which it has provoked in its readers from the date of its publication in 1928. Why? Is this anxiety due to the obscene, perhaps pornographic nature of the text? Ostensibly, *Histoire de l'œil* tells the story of the erotic and perverse games of a couple, the (male) narrator and his female accomplice, Simone, which become increasingly violent, leading to the death of a chaste friend, Marcelle, and then, after the witnessing of the enucleation of the bullfighter Granero in the arena of Madrid, the murder of a priest, and the cutting out of his eye. In the final scene of the narrative, the priest's eye is inserted into Simone's sex. This summary, however, gives absolutely no sense of the kind of text *Histoire de l'œil* is, nor does it account for the power of the text's obscenity, which affects the reader on every reading, or for the second part of the book 'Coincidences', in which the 'author' meditates on the links between the narrative and his biography. The text's questionable pornographic status is not the only factor determining its

---

[3] See P. ffrench, *The Time of Theory: A History of* Tel Quel (Oxford: Clarendon, 1995).
[4] See pp. 39–42.

transgressive force, and it is not what has made it a key text in the literature of modernity, exerting a hidden influence throughout the century, and not only in French writing and art.

Roland Barthes, in the seminal article 'La métaphore de l'œil'[5] (I will come back to the role this critical text plays in the history of the book's reading) gives an account of the text as the trajectory of the metaphor of the eye in its displacements along a chain of images, producing a transgressive eroticism through this displacement. The key moments of the text are thus provided by the juxtaposition of certain objects and the perverse uses to which they are put. The text begins with Simone sitting down in a bowl of milk, and the narrator looking at her 'chair "rose et noire" qui se rafraîchissait dans le lait blanc'.[6] This association of milk, which is white, and the saucer, which is round, 'returns' later in the text, when Simone and the narrator enjoy various erotic games with eggs (which are white, and globular), and then, more violently, when they witness the enucleation of the bullfighter Granero and Simone simultaneously inserts a castrated bull's testicle into her vagina. Both the eye and the testicle are whitish and globular. The text's final image is more violent still, as Simone inserts the enucleated eye of a priest into her vagina.

The force of the text lies undoubtedly in the transgressive eroticism it portrays, but the reader must also take into account the structural nature of this text, the way that the narrative seems generated not by any psychological depth in its characters, any motive, or any search for the resolution of an enigma, but by the structural permutations of a series of objects with each other and with the body of Simone.

My reading aims to account both for this structural aspect of the text, reading *Histoire de l'œil* through and after Bar-

---

[5] Roland Barthes, 'La métaphore de l'œil' in *Essais critiques* (Paris: Seuil, 1966). Barthes's article was originally published in the 1963 obituary issue of *Critique*, 195–6 (Aug.–Sept. 1963).
[6] G. Bataille, *Œuvres complètes*, vol. I, 14. Henceforward *O.C.* I. I refer to the 1928 version of the text. See pp. 34–37.

thes's critical article, and for its transgressive force, its affect. It also aims to give a reading of the text's position in literary history, for while the structuralist account of the text is by necessity synchronic, reading the text without reference to the context in which it emerged, it is my postulation that *Histoire de l'œil* also derives from a specific literary and aesthetic context, whose character I discuss. This context is specifically not that of Surrealism, the current which dominates 'avant-garde' literature and aesthetics in the 1920s, and the text needs to be dissociated from that context quite carefully, in terms both of its history and of its structural operation.

The Cut is deliberately not structured according to a rigid demarcation of chapters, treating different aspects of the text, its context, its themes, its intertexts. It operates, rather, according to a structuring which emerges from the process of reading, organized arbitrarily according to a series of key terms, and numbered. The numbering suggests the impossibility of closure for this reading, the impossibility of an exhaustive account. At the same time, a general movement may be ascertained, from a consideration of the problem of reading the text and the problems its reading has posed, to a discussion of the text's position in literary history, to a more detailed reading of certain elements and figures within the text.

1. / the cut

The history of avant-garde cinema is scattered with images of aggression against vision: visual shocks, violence done to the apparatus of projection, holes torn in the screen, dislocated frames, punctured celluloid. The radicality of an underground cinema which, through an apparently formalist refusal of narrative, brings us into proximity with the pure apparatus of the visual, the pure act of seeing, engenders a representation of itself as violence. This violence is figured as violence against the imaginary body of the spectator, as, for example, in Brakhage's *The Act of Seeing with One's Own Eyes*,[1] or, in a more abstract sense, Warhol's extended projections[2] or Snow's fixed camera rotations.[3] One might postulate, then, that this cinema of violence against vision finds its own mirror-image in a celebrated moment from early avant-garde cinema history: the image of the eye slit by a razor in Buñuel and Dalí's *Un chien andalou*, which, made in 1928, is roughly contemporary with Bataille's *Histoire de l'œil*.[4]

[1] A silent, hand-held camera 16mm film of an autopsy in Pittsburgh morgue which I saw in the season curated by Cerith Wyn Evans at the Institute of Contemporary Arts in London.

[2] For example, *Empire State Building* or *Kiss* or *Sleep* (1963).

[3] *Wavelength*, for example.

[4] Through curious happenstance, my own book is more or less contemporaneous with a recent film by Jo-Ann Kaplan, titled *Story of I* (produced by the British Film Institute), which announces itself to be loosely inspired by Bataille's *Histoire de l'œil*. Rather than a straightforward representation of 'what happens' in the text, the film produces a visual commentary upon it: it anchors itself in a scene of reading (a woman reads the text while in the bath) and produces, through animation and collage, images correlating to the reading of selected phrases from the text which suggest the aesthetic context of its time — interweaving images from Dalí and Buñuel's film, Bellmer and Man Ray's photography and others too numerous to mention, or unidentified by this viewer. It also exploits the structural operation of the text (see later) both in its explicit images (an egg sliced through horizontally, for example) and in its practice of montage. Jo-Ann Kaplan, who narrates the

I start with this incursion into the history of avant-garde cinema in order to introduce the theme of violence against the visual, and to use *Un chien andalou* as a way to point to the operation of the *cut* in Bataille's writing, part of a structural aggression against the coherence and security of the body of the spectator, the subject of speculation, the reader of the text.

In the article 'Œil' in the journal *Documents*,[5] written a few years after *Histoire de l'œil*, Bataille begins with a reminder of the association of the eye with seduction: 'Il semble, en effet, impossible au sujet de l'œil de prononcer un autre mot que séduction . . .'[6] Seduction: part of a dialectic of looks in which the subject recognizes their desire in the eye of the other; a structure in which the eye is the organ of desire, the locus of identity, the centre. But, Bataille adds: 'la séduction extrême est probablement à la limite de l'horreur'.[7] *A la limite*: on the other side of, or on the line of, this limit, seduction tips into horror. It is the operation of this passage over the limit, transgression, which Bataille's text performs, and of which the image I have cited is a symptom. Transgression ruins the dialectic of desire and recognition with the eye at its centre, as its pole, and inaugurates a different kind of structural movement, one of displacement and dislocation. The eye, at the limit, thus figures the operation of the transgressive cut which dislocates the eye as privileged locus of the structure of vision:

> A cet égard, l'œil pourrait être approché du *tranchant*, dont l'aspect provoque également des réactions aiguës et contradictoires: c'est là ce qu'ont dû affreusement et obscurément éprouver les auteurs du *Chien andalou* lorsque aux premières

---

film herself, also wisely uses her own translation. The film does not refer to the second part of the text, 'Coïncidences', however, restricting itself to the narrative.

[5] *O.C.* I, 187–90.
[6] Ibid., 187.
[7] Ibid.

images du film ils ont décidé des amours ensanglantés de ces deux êtres. Qu'un rasoir tranche à vif l'œil éblouissant d'une femme jeune et charmante, c'est ce qu'aurait admiré jusqu'à la déraison un jeune homme qu'un petit chat couché regardait et qui tenant, par hasard, dans sa main, une cuiller à café, eut tout à coup envie de prendre un œil dans la cuiller.[8]

What is happening in the disturbing final sentence of this paragraph? The film image produces, through some strange operation, an unfixing of objects from their usual function— the *petit chat couché* and the *cuiller à café*, whose subsequent juxtaposition then produces the transgressive image of the *œil dans la cuiller*. The contiguity of shape of the eye and the spoon, both ellipsoid and in some sense fluid, produced through the flattening or 'melting' of the circle, has seemed to demand their juxtaposition, just as the desirous lover, in *Un chien andalou*, watches the sliver of cloud pass across the moon and is suddenly, as if by chance, drawn to slit the eye of the woman he desires, given that he happens to have a razor in his hand. The link between the two images is the syntactically complex: *qu'un rasoir tranche à vif... c'est ce qu'aurait admiré...*, a copula whose logic is not that of cause and effect and whose seeming arbitrariness points to the effect of the text in engineering a transgressive, horizontal slippage from one image to another.

Dalí and Buñuel's image shows the operation whereby the eye of desire is cut *horizontally* across, and this horizontal cut then becomes part of a structural play of images; it is continuous, in the film, with the horizontal movement of a cloud across the moon and is hinged, so to speak, into this image through the filmic succession of images. In Bataille's article, the *cut* also enables a transgressive displacement across a textual surface which generates startling associations. The image from *Un chien andalou* not only functions because of its content, but also has a formal role within Bataille's article, inaugurating the operation of a transgressive structural play.

[8] Ibid. 187–8.

The cut is an image of extreme affective violence, an aggression against the body of vision, our body, *and* a strategic operation whereby symbolic space is (re)organized.

Technically, the filmic cut is that interstice between shots which itself is not seen, the instant which interrupts vision and remains invisible, but which enables the *montage* of images and the production of meaning. The cut itself is instantaneous, a negation of time which punctures it as substance and duration. The instant, the cut in time, functions, nevertheless, as the operation which inaugurates a structural play. *Histoire de l'œil* brings the moment of that instantaneous cut into the foreground and shows how it restructures and destructures meaning.

2. / obtuse

In the history of avant-garde film and film theory it is perhaps Sergei Eisenstein who most significantly champions the operation of *montage*, the production of sense through juxtaposition around the filmic cut or splice. In his recent book *La ressemblance des formes ou le gai savoir visuel selon Georges Bataille* the philosopher and historian of art Georges Didi-Hubermann[1] has signaled the proximity of Eisenstein's practice to that of Bataille in the *Documents* 'Dictionnaire critique', underlining that both are dialectical operations, but that they do not result in a third term as a resolution or higher term, and function instead as a *symptom* (see pp. 28–9). The symptom, in this sense, is a term which undermines and destructures the opposition; it is an object, a word, or a thing, which results from an unconscious sense of contiguity rupturing the surface of consciousness. The fixity of categories and objects, the stability

[1] Georges Didi-Hubermann, *La ressemblance des formes ou le gai savoir visuel selon Georges Bataille* (Paris: Macula, 1995).

of the opposition is destructured, unfixed, but *it is not dissolved*. In other words, this is not a vertical operation signaling a higher or lower reality, but a structural, horizontal play, an operation of destructuring along the surface of discourse. Bataille and Eisenstein's practice thus appears as a transgressive dislocation and destabilization of the order of reality. It is an operation of decoding which does not propose an alternative version but remains at the *horizontal* level of this structural undoing. It is something like the way Roland Barthes defines the activity of *reading*, when it is applied to the writerly text (*le texte scriptible*).[2] I want to approach my own reading of *Histoire de l'œil* via a discussion of Barthes in relation to Bataille.

Fortuitously, as Didi-Hubermann points out, the first to have noted the proximity of Bataille and Eisenstein is: Roland Barthes, in the article 'Le troisième sens'.[3] The impetus of this text relies upon the distinction which Barthes introduces between *le sens obvie* and *le sens obtus*, the first referring to the signification of the image, its symbolism or its evident meaning, the second referring to some element of the image which undoes the categories and oppositions which inform it, without dissolving them. The obtuse sense is a sense 'en trop', a meaning in excess, and has a materiality which overflows the boundaries of structured signification: 'l'émoussement d'un sens trop clair, trop violent'.[4] The obtuse sense is not ascribable to any purely aesthetic effect, indeed it 'includes' the beautiful and its opposite, but also undoes this opposition by introducing 'le dehors même de la contrariété'.[5] It is, as if to complete the Bataillean register of Barthes's terms, 'de la race des jeux de mots, des bouffoneries, des *dépenses inutiles*'.[6] Excessive, grossly material, a third term which falls outside the oppo-

[2] Roland Barthes, *S/Z* (Paris: Seuil, 1970), 10.
[3] Roland Barthes, 'Le troisième sens: notes de recherche sur quelques photographies de S. M. Eisenstein', in *L'obvie et l'obtus* (Paris: Seuil, 1982), 43–61.
[4] Ibid., 45.
[5] Ibid., 53.
[6] Ibid., 46.

sition *beautiful/ugly*, the obtuse sense is the equivalent of the operation of the cut which Bataille's texts execute. Barthes draws attention to this in pointing to the text 'Le gros orteil' (the title of which he *obtusely* recalls as 'Le gros orteil de la reine'[7]) by Bataille, from *Documents*, a text he will pick up and analyse later in his article 'Les sorties du texte', as 'pour moi l'une des régions possibles du sens obtus'.[8] The obtuse sense, then, is that element which sticks out (*qui sort*), which cannot be contained within the order of (any) discourse, which sticks out *for me*, and which sticks out because of a quality of irrecuperable materiality: the big toe, the curve of the headscarf of the weeping woman (in an image from Eisenstein's *La ligne générale* which is Barthes's main focus), the *punctum* of the later *La chambre claire*.[9]

3. / structural(ist)

The transgressive cut is an operation; it engenders a structural play, or a destructuring. Why this emphasis on structure, on the structural quality of what Bataille's writing does? Bataille might seem to be a writer whose texts would necessarily fall outside the sober, austere domain of that structuralism which dominated the 1960s, but which surreptitiously makes its return here.

Barthes is, of course, the first to have emphasized the structural quality of Bataille's transgressive eroticism in *Histoire de l'œil* in the 1963 article 'La métaphore de l'œil' which I have already cited. Barthes's article inaugurates the consideration of *Histoire de l'œil* as a *text*, as something other than a piece of surrealistic pornography. I hold that one cannot read *Histoire*

[7] Ibid., 53.
[8] Ibid.
[9] Roland Barthes, *La chambre claire* (Paris: Gallimard, Seuil, 1981).

de l'œil except in relation to Barthes, today. His account proposes the following: *Histoire de l'œil* is not the story of its characters, but the story of those displacements that occur across a chain of signifiers beginning with the eye, and their permutative intersections with each other. As a result, *Histoire de l'œil* has no depth, is a text which is all surface, which functions horizontally rather than vertically, and which, moreover, does not appeal, in its eroticism, to any basis in sex. It is the transgressive movement of displacement and association which in itself is erotic. The eroticism of *Histoire de l'œil*, then, is not expressive or representational, but structural, consisting in the play of associations which transgress the 'usual' functions and uses of words and objects, most significantly parts of the body.

What I want to emphasize in Barthes's account is how the structural play of association is itself transgressive of the closure of structure as such; how, in other words, this structural play is also a destructuring. I want also to suggest how, in 1963, Barthes's account of *Histoire de l'œil* is characterized by a tension between the structural complexity and movement of the text and the fixed binary oppositions of structuralism. In later texts, *after* structuralism, this tension is worked through, producing a different kind of critical writing in Barthes's second text on Bataille, the 1972 article 'Les sorties du texte'.[1] Tracing the evolution of Barthes's critical voice in relation to Bataille can then prove illuminating with regard to my own approach to *Histoire de l'œil*.

It pays to follow closely the terms of Barthes's analysis in 'La métaphore de l'œil'. The text follows the *migration* of the eye, or the cycle of its avatars:

> son histoire est [. . .] celle d'une migration, le cycle des *avatars* (au sens propre) qu'il parcourt loin de son être originel, selon la

[1] Roland Barthes, 'Les sorties du texte' in P. Sollers (ed.), *Bataille* (Paris: 10/18, 1973), 49–73.

pente d'une certaine imagination qui le déforme sans cependant l'abandonner . . .[2]

Migration: the implication is that the movement will return to its source. Cycle: the implication is that that the movement is a circuit. Yet: 'pente', 'déformer' imply a non-cyclical, non-closed destructuring play of deformation. This ambiguity will remain latent, since Barthes continues, in what follows, to work according to the oppositional pairs dear to structuralist method. Thus, in his consideration of the generic status of *Histoire de l'œil* Barthes unwillingly, it seems, accords the text the name of 'poem', according to the opposition novel/poem; itself, as Barthes stipulates, informed by the powerful structuralist pairs of metaphor and metonymy, paradigm and syntagm, vertical and horizontal, 'que la linguistique nous a appris récemment à distinguer et à nommer'.[3] With this last comment Barthes marks the historical moment of his text, 1963, the year of the publication of his own *Eléments de sémiologie*,[4] the high point of that structuralism informed by Jakobsonian linguistics, pre-Kristeva, pre-Derrida. While the novel proceeds through aleatory combinations of real elements ('combinaisons aléatoires d'éléments réels'[5]), the poem moves entirely in an imaginary space, and exhaustively completes all possible combinations of a fixed set of variables. Refusing the aleatory and arbitrary quality of the novel, the poem is the only alternative, according to the binary opposition Barthes uses. The poem, then, proceeds 'par exploration exacte et complète d'éléments virtuels'.[6] And, since the oppositions Barthes sets up, novel:poem::metonymy:metaphor (::syntagm: paradigm), associate the poem with paradigm and metaphor, *Histoire de l'œil* is classified as 'pour l'essentiel une composition métaphorique'.[7] But a parenthetical aside throws suspicion on this

[2] Barthes, 'La métaphore de l'œil', 238.
[3] Ibid., 239.
[4] Roland Barthes, *Eléments de sémiologie* (Paris: Seuil, 1963).
[5] Barthes, 'La métaphore de l'œil', 239.
[6] Ibid.
[7] Ibid.

choice of one of two possible poles: ('on verra que la metonymie y intervient cependant par la suite'),[8] and it is this suspicion which, with hindsight, we might see cracking open the opposition and, as Barthes suggests later, making it 'limp'. For the moment, on the crest of the euphoria of classification which Jakobson's opposition produces, Barthes continues: the eye is *varied* across a number of *substitutive* objects, it is named as a *paradigm* which produces substitutes which are the same (all globular) and yet different (since *named* differently). The substitutes are 'révélés comme des états d'une même identité'.[9] An image of *Histoire de l'œil* is thus produced which postulates the eye as the primary figure generating a series of metaphors for itself along a chain of paradigmatic similarity. The 'stations' of the metaphor of the eye are thus determined, in Barthes's analysis, by the combination of two elements: globularity and whiteness. The 'sphère métaphorique'[10] of the text can be fully constituted ('pleinement constituée')[11] by a chain, joined at either end, which runs from the eye (globular and white, and beginning with the letters 'œ') to the egg (the same), to the saucer of milk (round and white) to the bull's testicle ('d'une blancheur nacrée',[12] spherical) and back to the eye:

> Ainsi se trouve pleinement constituée la sphère métaphorique dans laquelle se meut toute l'*Histoire de l'œil*, de l'assiette de lait du chat à l'énucléation de Granero et à la castration du taureau . . .[13]

'Pleinement constituée'? Not quite, for the text does not end with Granero's enucleation, but with the enucleation of the priest (whose eye resembles Marcelle's) and its insertion into Marcelle's sex, which seems to weep tears of urine. Can this spherical structure (affording Barthes the fortuitous (or perhaps not) possibility of a structural model that resembles

[8] Ibid.
[9] Ibid.
[10] Ibid., 240.
[11] Ibid.
[12] *O.C.* I, 54.
[13] Barthes, 'La métaphore de l'œil', 240.

the eye itself) account for all the associations and potential intersections of the text? Complicating his own model, Barthes introduces a secondary metaphorical chain, the avatars of liquid, which will be 'exchanged' or 'crossed' with the first and primary chain of the eye, such that the text's structure becomes determined by the necessity of an application of one chain of metaphors to the other, the crossing and intercutting of the cycle of the eye (globular and white) with that of liquid and all the variants of its verbalization ('toutes les variétés de l'*inonder*').[14] We thus have a structural principle through which the text is generated, and which determines what happens in the narrative, subject to the combinations of the metaphorical chain rather than the aleatory circumstance of the real: the text will enact all the potential combinations of a primary metaphorical chain (that of the eye) with another (that of liquid) which is moreover associated with it.

Given this principle of cross-association, it is difficult to see what structural model may be constructed for the text's structural play. The difficulty may result from the inability to square the structural complexity introduced by the crossing or dislocation of one series of images with and by the other, with a structuralist model which is essentially tabular and exhaustive, with the structuralist demand for a fully constituted, exhaustive account. Or, can structural complexity be accounted for by a model of closure, *a* structure? What happens when a structural action (application, or the complication of one chain with the other) overflows the possibility of the table to account for it? The table turns into a infinitely extendible grid-like structure, the structure turns into a structuring, a moving process of structuration, such that the complete model itself can never be fully constituted. The displacement of the table, characterized by a fixed set of elements on one axis and on the other, towards the grid, characterized by the possibility of adding further elements, altering the components of the axis, undoing the stability of the axis, is symptomatic of the inca-

---

[14] Ibid., 140. Barthes's emphasis.

pacity of structuralist method to account for structural complexity and the *alteration* that Bataille's text practises.

What happens to the metaphor/metonymy couple here? It is made to limp;[15] a third term is added which undoes the stability of each term assured by the tension against its opposite. Barthes proposes that Bataille's eroticism is 'essentiellement métonymique',[16] characterized by the translation of meaning from one chain to the other. Metaphor/metonymy/metonymy. It is not metonymy as the opposite of metaphor which Barthes means here, but an operation 'within' metaphor which undoes it as a vertical paradigm, and undoes the opposition with metonymy. Metaphor always in the process of collapsing into metonymy . . . the vertical always falling to the horizontal . . . The metaphor or paradigm of the eye is not opposed to the syntagmatic and aleatory eye of the novel, but is undone as metaphor through its displacement into other elements which are not part of its original chain. The key moments of *Histoire de l'œil* are informed by this *alteration*, which Barthes terms a 'tremblement',[17] producing the syntagms 'casser un œil', 'crever un œuf', 'liquéfaction urinaire du ciel' . . . Paradigm/syntagm/displacement: a third term, outside the structure of opposition, comes to dislocate the binary pair: 'le terme de séduction *hors-la-loi*, (structurale)',[18] as Barthes will name it in the 1972 text. Not *either* paradigm or syntagm, but both simultaneously. The eye is *not* the egg, but at the same time, it *is* the egg ('tu vois l'œil? [. . .] C'est un œuf' says Simone).[19] It is the force of this transgressive copula 'is' which generates *Histoire de l'œil*, which is not the sign of equivalence nor of difference, but both at the same time.

[15] Barthes, 'Les sorties du texte': '[. . .] l'appareil du sens n'est pas détruit [. . .] mais il est *excentré*, rendu boiteux', 58 (Barthes's emphasis).
[16] Barthes, 'La métaphore de l'œil', 244.
[17] Ibid., '[. . .] le monde devient *trouble*, les propriétés ne sont plus divisées; s'écouler, sangloter, uriner, éjaculer forment un sens *tremblé* . . .' (244, Barthes's emphasis); 'Pour Bataille il s'agit de parcourir le tremblement de quelques objets . . .' (245).
[18] Barthes, 'Les sorties du texte', 58.
[19] *O.C.* I, 67.

What happens to structuralism between 1963 and 1972 which would account for the move from the difficulty I have proposed of maintaining Bataille's textual operation within a model of closure ('fully constituted') to the explicit recognition of a third term which dislocates the binary pair? What happens to Barthes's own critical approach between these dates? The response to this question would require a detailed analysis of the history of critical theory in France, beyond the scope of this study; for the sake of brevity it may, however, be encapsulated by a name, which may be Derrida, or Kristeva, or *Tel Quel*, whose work Barthes is receptive to, on his own admission.[20]

We might take the name Kristeva, and look at the way, in her early article on Barthes's *Système de la mode*, she recognizes the potential tautology of structuralist and semiological method (which was suggested by the way Barthes's critical model seemed to duplicate the sphericity of the images of the text), and proposes the introduction of process and dynamism.[21] However, given the importance of the notion of structure and my proposition that *Histoire de l'œil* does function structurally, it will be more useful to take the name Derrida, and to look briefly at a key text of 1966. In 'Structure, signe et jeu dans le discours des sciences humaines',[22] Derrida makes a decisive move in the dislocating of structure as a model of closure and the mounting (*montage?*) of a *structuring* as play. He writes:

> jusqu'à l'événement que je voudrais repérer, la structure, ou plutôt la structuralité de la structure, bien qu'elle ait toujours été à l'œuvre, s'est toujours trouvée neutralisée, réduite: par un geste qui consistait à lui donner un centre, à la rapporter à un point de présence, à une origine fixe. Ce centre avait pour fonction non seulement d'orienter et d'équilibrer, d'organiser la structure—on ne peut en effet penser une structure inor-

---

[20] Cf. Roland Barthes, *Roland Barthes par Roland Barthes* (Paris: Seuil, 1975), the table of 'intertexts' under the title 'Phases'.
[21] Julia Kristeva, 'Le sens et la mode' in *Séméiotiké* (Paris: Seuil, 1969).
[22] Jacques Derrida, 'Structure, signe et jeu dans le discours des sciences humaines' in *Ecriture et différence* (Paris: Seuil, 1967).

ganisée—mais de faire surtout que le principe d'organisation de la structure limite ce que nous pourrions appeler le *jeu* de la structure.[23]

And further on:

Le concept de structure centrée est en effet le concept d'un jeu fondé, constitué depuis une immobilité fondatrice et une certitude rassurante, elle-même soustraite au jeu.[24]

Derrida describes the 'event' to which he refers (ambiguously connected to the 'names' Freud, Nietzsche and Heidegger) as the beginning of an attempt to think the structurality of structure, and the centre not as a 'fixed locus', but as a 'function'. No tabular model anchored around a fixed term (or paradigm), Derrida's structure reveals it to have always been a differential movement, a play of displacement within which the notion of centre served as 'une sorte de non-lieu dans lequel se jouaient à l'infini des substitutions de signes.'[25] The centre or paradigm itself was subject to the play which it posed itself beyond.

This is 1966. In 1970, Barthes's *S/Z* opens with an emphasis on *structuration* rather than structure, production as opposed to product. Reading is not (now) the analysis of a text according to a pre-established, potentially universal, model (Barthes refers to the Buddhist notion of 'seeing a whole landscape in a bean'), but an open, potentially infinite, differential process. By now, also, the form of Barthes's writing has changed. 'La métaphore de l'œil' refused any recourse to psychology, wisely, given the material, and carefully refrained from any response to the obscenity of the text, insisting on 'une critique formelle'[26]—a formalist criticism sticking to the linguistic operations and categories of the text, already there in front of the reader and to be found. In *S/Z* Barthes puts himself in the position of the reader, becomes the subject-as-reader, producing therefore a mobile process or structuring,

---

[23] Derrida, 'Structure, signe . . . ', 409 (Derrida's emphasis).
[24] Ibid., 410.
[25] Ibid., 411.
[26] Barthes, 'La métaphore de l'œil', 241.

destructuring, restructuring, interrupted or 'starred' (*étoilé*)[27] by a self-conscious commentary on this reading. Structuring and/or destructuring, since the disarticulation of structure is the same movement of its articulation, its play.

In the 1972 text on Bataille, 'Les sorties du texte', given at a *Tel Quel* conference on Bataille, two years after *S/Z*, this internal structuring or disarticulation of the text and the organization of the critical text as a step-by-step reading has been replaced by a different kind of organization, without order (other than the arbitrary order of the alphabet), but fragmented into more or less short sections given specific names (eg. 'Déjouer', 'Paradigme', 'Orteil' etc.). The same structure will inform *Roland Barthes par Roland Barthes* and *Fragments d'un discours amoureux*. It is not a narrative structure or an exhaustive, structuralist table, but a mobile grid, which, like the writerly text, can be entered into at any point. Barthes refers to this organization as 'un ordre privé de sens'.[28]

4. / œ I

'Privé de sens'? perhaps, but the identification and naming of figures, their *choice*, that complex sub-titling which will become an element of Barthes's *style*, is it not the reintroduction of the subjectivity of the critic, not only as reader, but as subject of desire? *Le plaisir du texte*[1] had proposed the reader as a passive object of seduction for the text; the text desires the reader, endeavours to seduce him or her. The 'values' which star the critical text, the 'sorties' which Barthes names, are where the text sticks out 'for Barthes', where he is seduced by the text . . . crudely: where the text turns him on. If the

[27] Barthes, *S/Z*, 20.
[28] Barthes, 'Les sorties du texte', 49.
[1] Roland Barthes, *Le plaisir du texte* (Paris: Seuil, 1973).

subjectivity of the critic is foreclosed from 'la critique formelle', the subject adopting the neurotic (because characterized by a refusal of desire) position of metalanguage, the reintroduction of the subject as *reader* in *S/Z*, and then as subject of desire in 'Les sorties du texte' and the later work, is coincident with the rupture of the closed structuralist model and the move to structuring as play.

Le plaisir du texte, however, introduced the distinction between pleasure and *jouissance*, and showed that, while any text, even those of Bataille, had by necessity to have an element of desire, a coquettish aspect which was directed to the seduction of its reader, the *jouissance* of the text was something else, some other element which struck against the dialectic of desire and affected the subject in a different way. This 'other element' may be allied with the 'third term' discussed earlier, that 'sens obtus' which Barthes had recognized in the stills from Eisenstein films and whose similarity to Bataille's 'le gros orteil' he had noted. The third term, 'sens obtus' or *jouissance*, strikes the subject aslant, does not appeal to desire but to something that destroys the subject, annihilates him, tears him apart. What is it, then, about Bataille that strikes Barthes?

Is it that sense of compulsion which seems, behind the scenes, to have marked Barthes's choices as a critic from the beginning? The choice of Balzac's *Sarrasine*, we might remind ourselves, is determined by a comment made by Bataille in the preface to *Le bleu du ciel*, that it is among those texts which their authors must have felt compelled to write.[2] And Barthes admits in the discussion following 'Les sorties' that he was led to Michelet after a reading of Bataille's essay in *La littérature et le mal*.[3] Yet, in *Le plaisir du texte* Barthes hints at his distaste for Bataille's 'héroïsme insidieux'.[4] Indeed,

---

[2] Cf. *S/Z*, 23. Barthes reports that his choice was in the first instance influenced by an article by Jean Reboul in the journal *Cahiers pour l'analyse*, who took up the hint from Bataille's *Le bleu du ciel*.

[3] Barthes, 'Les sorties du texte', 70.

[4] Barthes, *Le plaisir du texte*, 50.

Bataille's virility would seem to jar with Barthes's increasing anti-hysteria and his passion for 'le neutre'. *Roland Barthes par Roland Barthes* gives a partial response:

> *Bataille, la peur*
>
> Bataille, en somme, me touche peu: qu'ai-je à faire avec le rire, la dévotion, la poésie, la violence? Qu'ai-je à dire du 'sacré', de l''impossible'?
>
> Cependant, il suffit que je fasse coïncider tout ce langage (étranger) avec un trouble qui a nom chez moi la *peur*, pour que Bataille me reconquière: tout ce qu'il écrit, alors, me décrit: ça colle.[5]

'Bataille, la peur'? Barthes's structural account of 'Le gros orteil', which he names 'Les sorties du texte', is conditioned, generated, compelled by fear, which in *Le plaisir du texte* is allied with *jouissance*.[6] Bataille produces that fear (that *jouissance*) which splits the subject apart, the 'trouble' which Barthes had already noted in 'La métaphore de l'oeil'.[7] The critical text, the reading therefore looks like a structural play with that element (obtuse, third term, *jouissance, punctum*) which wounds the subject as reader... Structural play *with* affect.

An imaginary title for this book could be Œ, figuring the displacement from closure (*O*), its slippage into horizontal movement (*E*), typographically representing the slit eye from *Un chien andalou*, pointing to the literal root of the text ('œil', 'œuf'), but also echoing Barthes's title *S/Z*. I proposed earlier that we cannot read Bataille's *Histoire de l'œil* without Barthes, and I want to extend that proposition to setting Barthes up as exemplary and tutelary reader, taking the example of the later texts and throwing it back on the object of Barthes's first article on Bataille of 1963. Barthes would be

---

[5] Barthes, *Roland Barthes par Roland Barthes*, 147.
[6] Barthes, *Le plaisir du texte*: 'Proximité (identité?) de la jouissance et de la peur', 77.
[7] Barthes, 'La métaphore de l'œil', 244. See p. 11, n. 17.

that strong critic, in Harold Bloom's sense, the anxiety of whose influence I am working through here.[8]

The cut, then, signals that dislocation of structure and binary opposition which inaugurates reading as structural play, but it also connotes the visceral obscenity of *Histoire de l'œil* which informs and generates any reading of it like a hidden or manifest *jouissance*.

5. / *informe*

A recent exhibition at the Centre Georges Pompidou in Paris, curated by Rosalind E. Krauss and Yve-Alain Bois had the title *L'informe: mode d'emploi*. It represented the culmination of much work done by Krauss, and others associated with the review *October*, around the term *informe* which titles one of Bataille's entries for the 'Dictionnaire critique' of the review *Documents*, which Bataille edited from 1929 to 1930. The *informe* was the name given to the transgressive operation of dislocation which characterized the perverse redefinitions of the dictionary; its own definition therefore designated the mode of action of the dictionary and its perverse and 'useless' (re)definitions of objects or ideas such as 'Figure humaine', 'Œil' (as we have seen), 'Chameau', 'Abattoir' and so on (as if this list could give rise to any logical continuation whatsoever). Using the *informe*, Krauss, in the exhibition's catalogue[1] and in her book *The Optical Unconscious*,[2] proposes a reformulation of the whole modernist project in the plastic arts. Where modernism seemed to move towards a sublime purity of form

---

[8] Cf. Harold Bloom, *The Anxiety of Influence* (Oxford: Oxford University Press, 1973).

[1] Rosalind E. Krauss and Yve-Alain Bois, *Formless: A User's Guide* (New York: Zone Books, 1997).

[2] Rosalind E. Krauss, *The Optical Unconscious* (Cambridge, Mass.: MIT, 1993).

Krauss insists upon the desublimating force of Bataille's *informe*, the operation of which undoes ('déjoue')[3] the pretension to ultimate abstraction. Krauss's project is possibly the most significant contemporary reading of Bataille, and has, moreover, proved productive of intense debate and reassessment of his work, and particularly the *Documents* period, in the field of art history. Georges Didi-Hubermann's *La ressemblance des formes ou le gai savoir visuel selon Georges Bataille* and Pierre Fédida's work on the relation of the *informe* to the symptom[4] are also part of this reassessment, which has rejuvenated the scholarly literature on Bataille, split between endless debates within the field of Continental philosophy over what Bataille does to Hegel, and accounts fascinated by the shock Bataille's virulence produces within the Academy. Krauss's reading of the *informe* is perhaps as seminal as Barthes's first text, and has the nature of a paradigm shift in the discourse which it represents.

If the label 'literature' sits uncomfortably upon *Histoire de l'œil*, it is nevertheless a text, to be read, and as such *Histoire de l'œil* is potentially only of tangential interest for the debate around the *informe*, focused as it is, primarily, on the visual field and its dislocations. Nevertheless, Krauss does consider *Histoire de l'œil*, although exclusively through Barthes's structural account of it, and uses it to point to the operation of dislocation, of slippage, which the *Documents* dictionary proposes. I want to reverse this polarity: while Krauss uses *Histoire de l'œil* as a literary parallel to the discursive operations upon the visual which *Documents* and the art in the recent exhibition effect, I will use her account of the *informe* to throw light upon my own reading of *Histoire de l'œil*. This

---

[3] Krauss uses the term 'Déjouer' as a chapter title in *The Optical Unconscious* (cf. 167–8, 184–5). She borrows it from Barthes, who uses it as the title of one of the 'fragments' or 'sorties' of 'Les sorties du texte', cf. 'Les sorties du texte': 53–4.

[4] Cf. Pierre Fédida, 'Le mouvement de l'*informe*', in *La part de l'œil* 10: 'Bataille et les arts plastiques' (1994), 21–8. See earlier for reference to Didi-Hubermann.

　　　　　　　　　　　　　　　　　　/ informe

will mean operating in the reverse direction to history, for the *Documents* project is, I think, informed by a structural operation which Bataille works out in earlier texts, in *Histoire de l'œil* and in *L'anus solaire*, written in 1927. *L'anus solaire* and *Histoire de l'œil* appear, then, as the textual space in which the operation of the *cut* is first played out.

In this light I want to pick up some key issues from Krauss's discussion of *Histoire de l'œil*, and from the work of Didi-Hubermann, and explore some of the ramifications of the theorization of the *informe* in relation to *Histoire de l'œil*. First, however, it will pay to see exactly how Bataille conceives of the *informe* in the *Documents* article. Here is the text in its entirety:

> *Informe*
> Un dictionnaire commencerait à partir du moment où il ne donnerait plus le sens mais les besognes des mots. Ainsi *informe* n'est pas seulement un adjectif ayant tel sens mais un terme servant à déclasser, exigeant généralement que chaque chose ait sa forme. Ce qu'il désigne n'a ses droits dans aucun sens et se fait écraser partout comme une araignée ou un ver de terre. Il faudrait en effet, pour que les hommes académiques soient contents, que l'univers prenne forme. La philosophie entière n'a pas d'autre bout: il s'agit de donner une redingote à ce qui est, une redingote mathématique. Par contre affirmer que l'univers ne ressemble à rien et n'est qu'*informe* revient à dire que l'univers est quelque chose comme un araignée ou un crachat.[5]

We have to begin, then, with a consideration of what words do, the tasks of words, not their meanings. What does the word *informe* do? To say what it means is to go against the direction of its task—to declassify, it would be to give it form, to give form to a word whose task is to 'unform'. *Informe*, then, is a task, and designates moreover other words as having tasks which do not square with their meaning, dislocate that meaning or spill out over it. Everything has its form, and philosophy insists on this, such that the *informe* gets crushed, gets stepped on like a spider or an earthworm, an insignifi-

[5] *O.C.* I, 217.

cance from the point of view of *le savoir*. What is the universe then? The *informe* ruins mimesis, ruins resemblance, the possibility of saying what the universe is 'like'; to say that it is *informe* amounts to saying that it ruins *le savoir*, its significance (meaning) is nothing, like spit, but what it does is to put a blotch on to the page of philosophy which screws up its pretension to define. The *informe*, in my reading, designates an operation that resists the tendency to ascribe meaning, to interpret and to fix. Its task is to interrupt meaning, inscribe a cut, which ruins the possibility of form (outline, circumscription—like the spider, spit or the earthworm).

The characteristics of the *informe* as Krauss conceives it are mapped out consistently across a number of different spaces. I can only give a very schematic account of them here, via Krauss's discussion of two artists: Duchamp and Giacometti. In *The Optical Unconscious*, in a consideration of Duchamp's revolving discs, Krauss designates as *informe*, formlessness, the sense of threat within an oscillatory rhythm of dissolution, of a lack of form: 'a threat carried by the very metamorphic rhythm itself, as its constant thrusting of the form into a state of dissolve brings on the experience of formlessness, seeming to overwhelm the once-bounded object with the condition of the *informe*'.[6] This threat of dissolution is not a fear of the void, of *le néant*, but a condition of the interstice between forms, the invisible *cut* which marks and produces their difference. It is something like the experience of *vertige* one has in reading *Histoire de l'œil*, as form gives way to form, another form, another... In other words, the *informe* is not nothing, not a substantial nothingness, a lack of form, but a characteristic of the movement of slippage, of difference. We are not thinking of presence and absence here, but of the very movement of becoming present as always conditioned by an operation of re-presentation, of difference.

At a number of points in *The Optical Unconscious*, Krauss discusses the piece *Suspended Ball* by Giacometti as 'a text-

[6] Krauss, *The Optical Unconscious*, 137.

book case of the *informe* as that was developed by Bataille.'[7] The key term here is *alteration*; as the slitted wedge swings over the ball a fundamental ambivalence is created whereby 'it is not clear, it will never be clear whether the gesture is a caress or a cut... whether the wedge is passively receiving stimulation from the sphere, or, passive-aggressive, is violating the surface of the ball...'[8] This alteration or ambivalence confirms the interpretation of the *informe* as not the absence of form, but the movement of its differentiation. One cannot say therefore that the *informe IS...* anything, unless the copula, the *is*, is taken as the sign not of equivalence but of a slippage. The eye *is* the egg. The eye *is* the sun. Krauss's version of the *informe* is definitively *structural*, emphasizing *alteration*, and to this extent her reading of *Histoire de l'œil*, which she briefly considers, refuses the definition of the *informe* as a thing, a substance.

Bataille's work is not, therefore, about an opposition between the 'high' and the 'low', about an exploitation of the ignoble versus the noble. In *Documents* the operation of the *informe* is brought to bear on diverse manifestations of human culture; the dislocation of the human more often than not takes the form of a foregrounding of the animalistic, not in order to propose an animal nature, a ground for the human in animality, but simply in order to depose the human from its sublime position. Krauss writes: 'In order to knock meaning off its pedestal, to bring it down in the world, to deliver it a low blow'.[9] The operation of the *informe* is a reminder of the body, of the *low* ('le bas'), but again, not in order to propose a primary physicality or sexuality, but for the purposes of desublimation. The body is not thereby sublimated into an origin, nor, in Krauss's conception, reified as an extra-structural, extra-cultural *thing*. In these terms, Krauss strongly disagrees, and I partially concur with her critique, with Kris-

---

[7] Ibid., 166.
[8] Ibid.
[9] Ibid., 157.

teva's apparent association of Bataille with the *abject* in the book *Pouvoirs de l'horreur*, seeing it as a recuperation of the *informe* through making it equivalent to the referent, the *thing*.[10] The temptation to read the *informe* entry cited above as proposing the irrecuperable materiality of the referent— *spit, spider, earthworm*—is strong, but one must insist on the effect (the task) the *informe* performs upon meaning, upon the idea, or upon idealism. It does not define a base materiality in opposition to the domain of ideas, or cultural forms, but poses materiality as that which undoes idealism. Not materialism/idealism, but the operation of materiality upon the ideal. A third term, which is not one.

In this light *Histoire de l'œil* is not about the body. It is not straight pornography, the staging of acts of copulation between bodies, but about the desublimation of the body as idea, as form. One should not talk about the body in relation to *Histoire de l'œil*, but about its parts, in the sense that Deleuze and Guattari talk of *desiring machines*,[11] an eye that has its own mind, a rhythm of insertion and ejection, a vagina that deploys itself around various objects. *Histoire de l'œil* realizes, in fact, the figure of the eye as an independent organ divested of the corpus in which it played such a sublime role. But this is not about a reification of the eye as an object, or the vagina as a thing; it is about the operation of the *cut* which severs the organ from the coherent body, a play around that severance. Thus we always have, in *Histoire de l'œil*, an eye which sees (the narrator) and a body (Simone's) around, on to and into which these part-objects are played.

Krauss's critique of Kristeva and the notion of the abject is continued in a round-table discussion in the review *October* on 'The politics of the signifier', which may also prove useful for my reading of *Histoire de l'œil*. Krauss reiterates her posi-

---

[10] This critique is developed in a discussion published in the journal *October*: 'The politics of the signifier II: a conversation on the *informe* and the *abject*', *October* 67 (Winter 1994), about which more later, and in Krauss and Bois, *Formless*.

[11] Cf. Gilles Deleuze and Félix Guattari, *Anti-Œdipe* (Paris: Minuit, 1972).

tion: 'I take the *informe* to be structural.'[12] She disagrees specifically with Benjamin Buchloh, who wants, it seems, to give the referent some role in the operation of Bataille's *informe*, to situate it historically. In effect, there is more than meets the eye in Kristeva's *Pouvoirs de l'horreur*, and Krauss's case seems rather overstated. The reference to Bataille in Kristeva's book is rather slim, consisting of a quotation on abjection as an unsuccessful exclusion of what must not appear in the constitution of the *human*. Kristeva points to Bataille as the sole writer to focus on the *weakness* of the law which institutes social order. The question of the law (*l'interdit*) and its strength or weakness is an eminently *structural* question, I think, hinging on the operation of transgression. What Krauss objects to is the tendency she identifies within Kristeva's approach, to reify, to fetichize *certain* objects or states as subject to exclusion. In effect, it seems that the object Kristeva sees as principally excluded, and threatening, abject, when this law of exclusion is not upheld, is the feminine, and particularly the maternal: 'Il s'agira donc dans ce qui suit de suggérer que cette relation archaïque à l'objet traduit en somme une relation à la *mère*.'[13] In this light Kristeva's notion of abjection should be linked to her earlier work on maternity,[14] as itself enacting the transgression of the law (of identity and unity), through the splitting of the mother and the emergence of the child as other in the field of the Symbolic. The main point to make here is that while Kristeva works with an essentially structural framework—inherited from the Bataillean notion of transgression—she uses it to address *certain* concerns: a reassessment of maternity and femininity. There *is* a tension between the structural aspect of Kristeva's work and the identification of the 'semiotic' as a 'pre-structural', 'real' space, substantified as maternal, feminine, and so on, but, as

[12] Krauss *et al.*, 'The politics of the signifier', 6.
[13] Julia Kristeva, *Pouvoirs de l'horreur: essai sur l'abjection* (Paris: Seuil, 1981), 79.
[14] Cf. Julia Kristeva, 'La joie de Giotto', 'Stabat mater', 'La maternité selon Giovanni Bellini', in *Polylogue* (Paris: Seuil, 1977).

Benjamin Buchloh points out in the round-table, these concerns are linked to their time, to history—they offer a way of addressing certain political stakes. Moreover, Kristeva nowhere suggests her reading of abjection as a reading of Bataille's work as such. What is more important to oppose, for Krauss, is the seduction of the *abject* for a reading of contemporary artistic practice—the reification of substance as abject in the work of artists such as Joseph Beuys, Louise Bourgeois, and others in their wake. The art history discourse on the abject, informed by Kristeva's psychoanalytic anthropology, threatens to draw Bataille into complicity with this reification and to forget the structural nature of the operation.[15]

Kristeva's apparent 'return to the referent' with *Pouvoirs de l'horreur*, may be seen, moreover, as part of a movement in the current of post-structuralist thought in France. Is not Barthes's final text *La chambre claire* a final recourse to denotation, to the irrecuperability of the real about which one can only say 'c'est ça',[16] after the failure of various attempts to outmanœuvre the insidious machinations of the Symbolic, of the *doxa?* Is there not a movement within the work of Lacan from work on the functioning of the Symbolic and on the lure of the Imaginary, towards a consideration of the repetitive thrust of the Real, its manifestation as a kind of death's head?[17] Return of the repressed? Perhaps so, but it does offer an antidote to the synchronic abstractions of structuralism and a

[15] For example, Frances Morris in the catalogue for the exhibition *Rites of Passage: Art for the End of the Century* (London: Tate Gallery, 1995) (which featured Beuys, Bourgeois, and an introductory interview with Kristeva) writing on the artist Mona Hatoum: 'Bataille's concept of "base materialism" is not so far from more recent notions of abjection, defined most famously by Julia Kristeva . . .', 104.

[16] Barthes, *La chambre claire*. See also Krauss and Bois, *Formless*, 192–3, and Diana Knight, *Space, Travel, Utopia* (Oxford: Clarendon, 1997), who stresses the return to *denotation* of *La chambre claire*.

[17] Lacan's commentary on Holbein's *The Ambassadors* in *Les quatres concepts fondamentaux de la psychanalyse* is what I have in mind here; the seminar from which it is taken took place a lot earlier, of course, 1964. Cf. Lacan, *Les quatre concepts fondamentaux de la psychanalyse* (Paris: Seuil, 1973).

/ informe

possibility of situating, historically and, let's say, erotically, the *action* of the structural, its *affect*. Bataille's appeal, *now*, may be partly to do with this 'return of the real', a desire to dirty one's hands, to 'get down into the shit', as an eminent art historian has put it.[18] Krauss's emphasis on the *structural* nature of Bataille's early work on the culture of his time is to this extent a welcome check to a potentially exploitative and redundant discourse.

Formless but not abject, then? The *October* discussion comes to the conclusion of an alternative position, a compromise position in some sense, where the *informe* is characterized as structural *and* by a 'pragmatics', a 'situation'[19] in which its action is effective. In the context of my own reading of *Histoire de l'œil* I think this is a useful way of thinking about the operation of the text as structural but also producing an effect or an affect on the reader, which is not about feeling disgusted at the portrayal of certain bodily substances, but a sense of vertiginousness at their transformation into one another. The *affect* of structural play as a *vertige* which induces whatever physical manifestation—nausea, revulsion, annihilation, *jouissance*, terror, intellectual jubilation. The abject of structure is the absence of ground. This abject is and is not a thing. It is and is not the eye, the egg, for these objects are and are not themselves. The Bataillean object is abject to the extent that it is not what it is: leakage, spillage, overflow.

In the 'politics of the signifier' discussions Buchloh insists, as we saw above, on the *situated* aspect of the *informe*: 'the *informe* is both bodily and social. It's breaking rules, rupturing conventions *and* situating that rupture.'[20] Yve-Alain Bois puts it more forcefully: 'You have to spit accurately, in the right

---

[18] Griselda Pollock, in a discussion following a paper by Judith Still on Kristeva's use of Bataille in relation to abjection, at a conference held at the Institute of Romance Studies in 1996 on the work of Kristeva, organized by Anne-Marie Smith.

[19] Krauss *et al.*, 'The politics of the signifier', 20.

[20] Ibid.

soup.'[21] The necessity to situate the soup, so to speak, into which Bataille is spitting, with *Histoire de l'œil* is one I attempt to address in this book. If Krauss *et al.* refer in passing to *Histoire de l'œil* as a kind of matrix or blueprint for the operation of the *informe*, basing their account essentially on Barthes's structuralist reading of 1963, they do not address the historical specificity of the text, its situation.[22] While the articles of *Documents*, primarily relating to a visual material which accompanies them, immediately signal the cultural 'forms' they undo, it is not so evident what is being undone or transgressed in *Histoire de l'œil*. As a transgressive text one would need to situate the *law* which is being transgressed. This requires a work of literary analysis—of what constitutes 'literature' in 1928 (which can only be partially addressed here)—of how the informal operation of *Histoire de l'œil* relates to the narrative which is its framework—of how the text takes up and transforms certain literary figures—of how the 'Coïncidences' section which closes the text takes up and plays with the discourse of psychoanalysis. A work of *reading* of the textual discourses which *Histoire de l'œil* transgresses, while including them.

In this light the work of Georges Didi-Hubermann, congruent in many ways with the *October* project but significantly different in key respects, is relevant for my own work. Didi-Hubermann's project in his book *La ressemblance des formes* is an exhaustive analysis of Bataille's *Documents* articles, the way in which he reads the cultural forms of his time, directed towards a rethinking of the notion of resemblance, or mimesis, and thus of the whole basis of the aesthetic. The historical specificity of *Documents* is addressed in a way which Krauss *et al.*, concerned with a position-taking within art history and the discourse of contemporary art, do not account for. It will therefore be useful to address briefly some points

---

[21] Ibid.
[22] Krauss mistakenly gives the date of publication as 1926.

in Didi-Hubermann's account which are of immediate relevance to my own reading of *Histoire de l'œil*.

6. / symptom

The *Formless* catalogue signaled two points of contention with Didi-Hubermann. One was the notion of *ressemblance* which features in the title of his book. According to Yve-Alain Bois, with this term Didi-Hubermann 'reintroduces wholesale everything the concept of the *informe*, such as we understand it, wants to get rid of'.[1] The *informe* would be 'reduced to' a 'rhythmic condition of form'.[2] In my reading this would not be a reduction. The *informe* radically destabilizes the *ressemblance* which it qualifies. Didi-Hubermann's project is to show what the operation of the *informe* does to mimesis. To oppose the *informe* diametrically to resemblance, not to see it as an operation internal to the condition of form is to want to see it as the absence of form, as nothing, which, it seems to me, is a gesture Krauss and Bois also want to resist. Not to see the operation of the *informe*, as Didi-Hubermann does, as a 'rhythmic condition of form' is to miss the point about the nature of the operation itself. It is an operation of displacement within that of resemblance, of metonymy which collapses the verticality of metaphor from within, a copula which is the mark not of equivalence but of slippage.

The second point of contention is in many ways a corollary of the first. Bois disagrees with consideration of the *informe* as part of a dialectical movement. According to Bois: 'Didi-Hubermann incessantly makes the thinking of the *informe* into a dialectics, a dialectics aimed at the assumption of a third term, with the Hegelian synthesis neatly replaced by

[1] Krauss and Bois, *Formless*, 80.
[2] Ibid., 80, citing Didi-Hubermann.

"the symptom" '.[3] This refusal of the dialectic is to my mind at once a refusal of the historical specificity of the operation of the *informe*, a refusal of its engagement with precise forms, and a refusal of any effect of its structural operation. It remains to be seen, however: what kind of dialectic, what kind of symptom? A discussion of the symptom can lead us on directly to my own discussion of *Histoire de l'œil*, in the specificity of its historical and cultural context.

For Didi-Hubermann the movement of the *informe* is dialectical because it is not a simple negation or privation of form. And this dialectic is 'symptomal' (Didi-Hubermann chooses to use the term 'symptomale' rather than 'symptomatique' to suggest the 'formal, critical and intransitive' aspect of what he intends)[4] in that it is not devoted to synthesis (closure or logical reconciliation), but to the symptom. But what is the symptom? The meaning of the term in this context goes beyond its usual acceptance in clinical discourse. Didi-Hubermann refers to it as 'une catégorie critique d'ordre très général, dépassant et même subvertissant la signification clinique que ce mot a pris depuis le XIIIème siècle dans le champ très particulier des sciences médicales'.[5] I would propose the following account, drawn from my reading of Didi-Hubermann and in the context of *Histoire de l'œil*: the symptom is a kind

[3] Ibid., 69.

[4] Didi-Hubermann draws quite heavily here on the work of analyst (and writer) Pierre Fédida (cf. 'Le mouvement de l'informe') who, he notes, remarks on the distinction between the clinical use of the term *symptom* and its 'metapsychological' use. I will not look in detail at Fédida's work here or the extent to which Didi-Hubermann's use of the term *symptom* derives from his work. But it may be pertinent to note the comparison Fédida draws between the *informe* and the 'travail du rêve' in Freud's work, the formation of the image (as symptom) through a process of deformation (*Zerrbild*—deformation of the image), and between the *informe* and the 'associative derivations of the image' in Freud's account of the Wolf-Man (*La part de l'œil* 10, 23). It is also pertinent to note, as Didi-Hubermann and Fédida do, how all of Bataille's work of this period is produced *in relation* to Freud and psychoanalytic culture (from his analysis with Adrien Borel onwards), but not necessarily with reference to it. I qualify this note later, on p .160.

[5] Didi-Hubermann, *La ressemblance des formes*, 333–4.

of sign. It is the indication of a disorder, of a disease. But in the psychoanalytic sense it is also the sign of a conflict, the overcoming of a repression, of a crossing of boundaries, of a transgression. The symptom is also a manifestation, it is manifest, but it cannot be interpreted directly, it does not refer directly to the trauma or disease it results from, but has undergone a work of transformation, a work of displacement of forms. The *informe* as a symptomal dialectic thus describes an operation of the transgression, dislocation or rupture of form which produces a manifest sign which shows (*montre*) that process of dislocation but also results from it. Or: a form which would include within it the evidence of its own rupture. A form which is not one. Symptomal form, then, is a form which is always already ruptured, deformed, unformed, hollowed out by the movement of differentiation which 'informs' it. The maintenance of this notion of a form (not form) is evidently difficult—it is difficult not to just slip into a consideration of it as a *new* form, leaning too heavily on the positive side of the operation, or a consideration of it as nothing, overemphasizing the negative turn. The difficulty of form (not form) whose contours are effaced by its own inkstain,[6] is connoted by the clinical register of the word symptom: pain, tension, illness.

For Didi-Hubermann the fact that the operation of the *informe* designates a symptomal dialectic or 'une économie symptomale des formes'[7] suggests first of all the impossibility of synthesis. Moreover, it suggests that the image or form is a symptom which bears within it the dislocation or transgression it has undergone or is undergoing. The symptom of seeing ('le symptôme du voir'), for example, would always be informed, in Didi-Hubermann's words, following Bataille, by the act of 'écarquiller les yeux, jusqu'à en crier'.[8] *Histoire de l'œil*, as

[6] See later, p. 33.
[7] Didi-Hubermann, *La ressemblance des formes*, 336.
[8] Ibid., 337, citing Bataille, 'Le gros orteil': 'Un retour à la réalité n'implique aucune acceptation nouvelle, mais cela veut dire qu'on est séduit bassement, sans transposition et jusqu'à en crier, en écarquillant les yeux: les écarquillant devant un gros orteil', *O.C.* I, 204.

symptomal text, would thus tell the story of the act of seeing already conditioned by a deformation of the eye, its violent dislocation. It would stage a movement of regression[9] (regression along the chain of displacements which led to the sublimation of the eye, its construction as definitive object of seduction) of the faculty of vision and thus an undermining of the whole edifice of 'speculative',[10] 'enlightened' human culture, towards the archaic (in a structural sense), 'base' uses of the eye, to which the eye can be put: baseness of the eye as 'friandise cannibale', or as viscous, round object ejected from its socket. *Histoire de l'œil* could be read as a narrative of this regression, in various phases: first, the eye not as that which sees but as that which is looked at[11] (whence the privileging of the fixed, fascinated stare in the text); second, the refusal of the eye as organ of knowledge and phenomenological orientation; third, the annihilation of the eye as 'window' to the soul of the living body (which we can see in the scene where Simone is haunted by the open eyes of the dead Marcelle— 'Les yeux ouverts de la morte'); fourth, the ejection of the eye from its socket (the death of Granero) and ultimately the eye displaced downwards to the sexual parts. In the final scene,

[9] Didi-Hubermann, following Pierre Fédida, comments on the relevance of the infantile to the 'dialectique régressive' of the *informe* (Didi-Hubermann, 250–1, Fédida, 23). In effect, the infantile qualities of the activities of the narrator and Simone in *Histoire de l'œil*, beginning with the episode of the saucer of milk, are difficult to ignore. Cf. also Bataille, in the 'Préface à l'*Histoire de l'œil*' in *Le Petit* of 1943: 'W.-C. était lugubre, autant qu'*Histoire de l'œil* est juvénile', *O.C.* III, 59. One would evidently want to follow through, if space allowed, the resonances of this *infantile regression* with Freud's 'polymorphous perversity'.

[10] The deconstruction of the notion of 'speculation' as transcendental signifier of the 'phallogocentric' regime has of course been carried out by Luce Irigaray, in her *Speculum de l'autre femme* (Paris: Minuit, 1974). See also Jean-Luc Nancy, *La remarque spéculative* (Auvers-sur-Oise: Galilée, 1973).

[11] There is much to say on this notion of 'being looked at'. Cf., among others, Didi-Hubermann's aptly titled *Ce que nous voyons, ce qui nous regarde* (Paris: Minuit, 1992), and Michael J. Popowski, 'On the eye of illegibility: legibility in *Histoire de l'œil*' in Leslie-Ann Boldt (ed.), *On Bataille* (Albany N.Y.: State University of New York Press, 1995).

/ symptom

in fact, the eye returns to its position as 'that which sees'—the narrator describes the priest's eye, which looks like Marcelle's, looking at him from within Simone's vagina. But this seeing eye bears the traces of its transfigurations, its displacements; it is a dead, emasculated eye re-located in a different hole.

In this light *Histoire de l'œil* appears as more than 'simply' a matrix for the operation of the *informe*, or an exemplification of it, although the latter and the debate surrounding it can provide useful pointers to its functioning. The text would inscribe a violent rupture in the history and structure of the *act of seeing*; it would narrate the deformations to which the eye, privileged organ of human knowledge and perception, is subject.[12] If the central position of the eye and its culture is partly what defines the 'human', *Histoire de l'œil* would narrate the transgression of the law which marks off the eye from its carnality, from its qualities as an object. For the objectal qualities of the eye are occluded in the act of seeing, or in the consideration of the eye as window or aperture, or surface upon which images are projected. In order to see, the

[12] It would be too easy to locate *Histoire de l'œil* as an Oedipal drama, its transgression a self-mutilation (on the part of the biographical subject— Georges Bataille) induced by a castration anxiety arising from the sight of the mother's genitalia. This seems to me partly the impetus of Susan Suleiman's article 'Transgression and the avant-garde: Bataille's *Histoire de l'œil*' in *Subversive Intent* (Cambridge, Mass.: Harvard University Press, 1990). Suleiman writes: 'As far as Bataille's text is concerned, it is clear that whichever interpretation one emphasizes, the focus is on the son's view of the mother's genitals, which invariably leads him to a recognition of sexual difference and to a split in his own experience . . .' (86). Transgression, in Bataille's schema, is a generalized, structural operation which would resist closure within Freud's account of the Oedipus complex. This is not to say that Bataille does not *play* with the imaginary possibilities offered by the mother/father/infant triad, in 'Coïncidences'. A proposition that the eye which is dislocated, 'touché', in *Histoire de l'œil* is the eye of the father would be justified by the account in 'Coïncidences' of the traumatic origin of the *récit* in the vision of the 'white' eyes of the blind and syphilitic father as he urinates (see also p. 175, note 3). But this reading would have to account for the way in which Bataille pre-empts any 'innocent' psychoanalytic or biographical account through its inclusion as 'already' part of the fiction. See p. 157.

qualities of the eye as viscous, white[13] and ellipsoid have to be forgotten; they are thus reactivated in this transgressive narration.

Histoire de l'œil is thus a story of the transgression of the taboo: 'il ne faut pas toucher à l'œil'.[14] The 'story of the eye' is the story of an imaginary regression of the eye along a chain of displacements from its sublimated position within the corpus of the *human*. The task of the word 'œil' is thus to put to work those qualities occluded by the sublimation of the eye as seeing organ of knowledge and thereby to undo all the cultural and philosophical architecture constructed around sight.[15]

'Toucher à l'œil': the taboo whose transgression *Histoire de l'œil* narrates passes through various phases; it is 'regardé', 'touché', 'bu', 'tranché', 'coupé'. 'Toucher à l'œil'[16] is also an operation which paradoxically represents the deformation of the very act on which depends this representation—the act of reading. It narrates the regressive desublimation of the position of the subject of vision, and therefore of the subject as reader. The *reading* of *Histoire de l'œil* would thus correspond to an image of vision mutilated—the cut of the eye which the infamous image of *Un chien andalou* proposes.

---

[13] See later, p. 94. The white of the eye is of course the part of the eye surrounding the iris and pupil. When the eye is turned upwards, into the skull, the iris and pupil disappear. If the eye is described as white, the apparatus of seeing is thus denied.

[14] Cf. Didi-Hubermann, *La ressemblance des formes*, 75.

[15] As a 'symptomatology' of cultural forms which are predominantly visual, the *Documents* period, from 1929, is thus preceded 'logically', or structurally, by *Histoire de l'œil*, which installs a prefatory desublimation of the visual, and thus of 'art', a category into which *Documents* threatens to fall.

[16] A powerful echo imposes itself here of an article by J. Derrida on the work of Jean-Luc Nancy, 'Le toucher', in *Paragraph*, 16:2 (July 1993), in which the figure of a touch upon the eyes is articulated as a violent aporia.

7. / spilled ink

When first published, Bataille's *Histoire de l'œil* is said to have 'circulated' among the intellectual and artistic milieux of Paris.[1] If the objects within the book can also be said to circulate, as Barthes's identification of a 'sphère métaphorique' would suggest, this circulation is nevertheless broken by a tendency towards rupture, towards spillage. The sun is 'liquefied', bodily fluids inevitably flow, jet, inundate and drench. Taking this metaphor further, the 'circulation' of the book among its readers, the economy of its exchange, is interrupted by its *dépense*, its overflow. In this context Jean-Luc Nancy has written convincingly of the 'exscription' of Bataille's writing, 'exscription' being that kind of writing which spills out of itself to touch the real, the *corpus*:

> ... Bataille immediately communicates to me that pain and that pleasure which result from the impossibility of communicating anything at all without touching the limit where all meaning spills out of itself like a simple ink stain on a word, on the word 'meaning'. This spilling and this ink are the ruin of theories of 'communication', conventional chatter which promotes reasonable exchange and does nothing but obscure violence, treachery and lies, while leaving the power of unreason no chance of being measured.[2]

Nancy's 'spilling', which ruins 'reasonable exchange', suggests a radically aneconomic principle of *dépense*, expenditure. The spillage of Bataille's writing would be that aspect of it which overflows and ruptures the capacity of words to be exchanged as meaning in a contractual communication. There would be something irreducible about Bataille's writing, something of

---

[1] One hundred and thirty-four copies of the book were printed, illustrated by eight lithographs by André Masson, and published by René Bonnel, who in the same year brought out the anonymous *Le con d'Irène* (whose actual author was Louis Aragon).

[2] Jean-Luc Nancy, 'Exscription', in *Yale French Studies* 78, *On Bataille* (Yale, 1990), 47.

the nature of an inkstain, a formless blot. Nancy's term 'exscription' translates that irreducibly corporeal aspect of Bataille's writing which produces an effect, pleasure or pain, on the body of the reader, on what we might propose as the 'Real' of reading. The word *dépense*, then, has the advantage of operating both in the register of the body, of its fluids and their spillage, and in the register of the heretical economics which Bataille will study later in *La notion de dépense* and *La part maudite*,[3] and this implies a problematization of the circulation of the book, of *Histoire de l'œil*, as an object, its exchange-value.

As an object the book has a strange history. It is currently available in two versions, the first version of 1928, and a subsequent version first published in 1940,[4] with alterations by Bataille. Both versions, that of 1928 and that of 1940, were published with the author's name given as 'Lord Auch'. Since 1926 Bataille had been writing articles on numismatics under his own name, Georges Bataille, in the review *Aréthuse*, and previously to *Histoire de l'œil* had written his first work, *Notre Dame de Rheims* and the short *L'anus solaire*; his first publication in book form, under the name Georges Bataille, would not be until 1942 with *L'expérience intérieure*; he took care to separate, nominally at least, the identity of Lord Auch from that of Georges Bataille.

While the first edition, illustrated by artist and friend André Masson, was published in Paris, a re-edition of the second version was accompanied by eleven illustrations by the Surrealist artist Hans Bellmer, with the place of publication given as Seville. Why this displacement, outside France? We might propose a desire to escape prosecution for obscenity, but the re-location of the text in Seville is also part of a consideration of place and displacement. The *récit* ends at the 'church of Don Juan' in Seville. Spain, in *Histoire de l'œil*, is

---

[3] Cf. *O.C.* I and VII.
[4] For both versions, see Bataille, *O.C.* I.

the place where things get acted out, throughout Bataille's fiction (*Le bleu du ciel* sets most of its action in Barcelona). Spain is also the country of *tauromachie*, of Don Juan.

To place is to locate. To name is to identify. To write about Bataille as author is to postulate his literary authority and identifiability. But Bataille is not the author of *Histoire de l'œil*. Lord Auch is the author of that text. A subsequent re-edition in 1967[5] of the second version of the text, by Pauvert, without Bellmer's illustrations, nevertheless identifies the author as Georges Bataille, and a re-edition of this by Pauvert in 1979, accompanied by the texts *Madame Edwarda* and *Le mort*, relegates mention of 'Lord Auch' to the small print.[6] The implication is that the reasons for which Georges Bataille used a pseudonym are now redundant, society having advanced in terms of the legality and tolerance of obscenity. There is, no doubt, incontrovertible evidence that this is in fact the case. But does not absolute availability, in the commercial market-place, carry with it a certain ideological charge? And in the context of Nancy's comments, wouldn't the book's exchange, in whatever quantity, tend to work against, or at least be resistant to, the more fundamental 'communication' which the text evidently aims at? Rather than preventing (legally) the reading of the text as such, what now occurs is the imposition of a 'certain reading', and a concomitant notion of authorial presence and accountability.

Among the versions of the text available, in spectacular fashion, for the contemporary reader, we can list: both versions of the text in the first volume of the *Œuvres complètes* of Georges Bataille, published in 1970 and prefaced by Michel Foucault; the Pauvert edition of 1967; the 10/18 *format de poche* edition which places it alongside *Madame Edwarda* and

---

[5] Georges Bataille, *Histoire de l'œil* (Paris: Pauvert, 1967).
[6] Georges Bataille, *Histoire de l'œil*, *Madame Edwarda* and *Le mort* (Paris: Pauvert, 1979).

Le mort;[7] and a further *poche* edition in Gallimard's *L'Imaginaire* collection.[8]

While the *Œuvres complètes* version obeys a logic of homogenization, in its will to collect all of Bataille's writings together, and in doing so irons out the very evident heterogeneity of the writings produced by that individual, it does add to the text a full bibliographical apparatus, and stipulates on its ninth page that *Histoire de l'œil* is 'par Lord Auch'. The other

---

[7] A re-edition of the 1979 Pauvert edition.

[8] The English translations of the text offer equally problematic representations of the text. For readers of the text in English, the only version available is Joachim Neugroschel's translation (of the first version of the text), first published by Urizen books in 1977, then by Marion Boyars in 1979, and subsequently by Penguin. A previous translation by Austryn Wainhouse under the title *A Tale of Satisfied Desire* appeared from Olympia Press in 1953. The Neugroschel translation, which became the current Penguin paperback edition in 1981, also includes two critical essays: Roland Barthes's 'The metaphor of the eye' and Susan Sontag's 'The pornographic imagination'. Sontag's essay appears for the first time in *Partisan Review* and then in her book *Styles of Radical Will*. While we might wish to celebrate the addition of critical, scholarly texts which presumably ward off attempts to exploit the text's lubricious appeal and which underline its literary status, this again is a strategy, on behalf of a certain kind of normativity, to impose a reading of the text, to circumscribe it, to deny the multiplicity of its readings. Again, a number of paratextual signs around this book effect a masking of the specifically textual activity of *Histoire de l'œil*, and its historical specificity. The text is designated (by Marion Boyars) as 'a major example of French Surrealist writing'; the back cover stipulates that it 'reveals the forbidden territory of adolescent sexuality and fantasy', both characterizations I would strongly reject. The translation of the text as *A Tale of Satisfied Desire*, by Austryn Wainhouse (his name is given as 'Audiart' on the book) gives the author's name as 'Pierre Angélique', the pseudonym used by Bataille for *Madame Edwarda*, also translated by Wainhouse as *The Naked Beast at Heaven's Gate*. Both translations originate from the circle of Anglo-American expatriates in Paris in the early 1950s, around Maurice Girodias's Olympia Press and George Plimpton's *Paris Review*. Bataille had himself given both texts to Girodias, who, under his real name, Maurice Kahane, was involved in the publication of the review *Critique* in 1946. Despite the title, which Michel Leiris justly criticizes for imposing the satisfaction of desire upon a text which through its strategy of formal displacement could have none, the Olympia Press translation is perhaps the most appropriate, since it maintains the unidentifiability of the author, and associates Bataille with Sade, whom Wainhouse also translated for Olympia Press. The translation, however, is beset by serious errors of interpretation.

editions, however, are guilty of the putative crime of the elision of the fictional name 'Lord Auch' and the attachment of the text to the socially identifiable and morally responsible *individual* Georges Bataille. The Gallimard *poche*, which reproduces the 1967 Pauvert edition, is especially crude in its presentation of the text, 'par procédé photomécanique', and seems to propose a patina of authenticity in its reproduction of Bataille's hand-writing at the end of the book, not actually the text of *Histoire de l'œil* itself, but of the 'Plan d'une suite', which is also appended. Here the individualization of the author is carried to the extent of identifying a gesturality attached to it, thus identifying a body, an origin, and what is effectively denied by this process of individualization is the autonomous effect of the text. This repression of text perniciously allows an identi-fication of the potentially pathological body of the writer who produced the work, who signs his name to it: Georges Bataille.

As Michel Foucault has powerfully argued in his history of sexuality,[9] 'sexuality' itself is a discourse which covers and is determined by a certain conception of the individual. It cements and justifies the notion of the morally and socially responsible individual who is the locus of a 'sexuality' itself anchored in the ideologically determined notion of 'desire'. The identification of the individual, rendered socially responsible for his acts (his texts) through the 'name of the author', as the site of a natural or deviant sexuality, is a way towards the individualization, pathologization and therefore the limi-ting and enclosure, of a transgressive thrust which precisely works against that limit, wants to break it open. In Bataille's own *L'érotisme*, the Christian era, roughly corresponding to the period Foucault determines as that of 'sexuality', sees the division of 'le sacré faste' and 'le sacré néfaste', or pure and impure transgression.[10] While the transcendent and invisible God is the pole of a decorporealized, sanitized transgression, 'impure' transgression, that relative to the body, becomes rel-

---

[9] Michel Foucault, *L'usage des plaisirs* (Paris: Gallimard, 1976).
[10] See *O.C.* X.

egated to the lower orders, to evil. Eroticism thus becomes associated with evil, and the current psychological or psychoanalytic conception of eroticism as pathology (despite Freud's insistence on the objective reality of 'polymorphous perversion') is a variant of this. Bataille, identified as authorial body, becomes the 'sick' or 'perverse' individual responsible for the 'pornographic', 'dirty' book *Histoire de l'œil*, and this identification closes the text within the ideologically determined field of 'sexuality'.

But I am using the name Georges Bataille, and will continue to do so, to identify the author of *Histoire de l'œil*, and to identify the author of an entire œuvre. It is important to refer to the writer of *L'expérience intérieure*, *L'anus solaire* and *Manet*, as the same as the writer of *Histoire de l'œil*. It is also important to recognize the strict correspondence between the writer of the *récit* of *Histoire de l'œil* and the writer of its second part, 'Coïncidences', which reflects on the experience of its 'author', and his personal obsessions. What must, however, be guarded against is an isolation of 'Bataille' as a pathologized individual whose imagination exists somehow outside the literary and social context of his time, whose sexual obsessions can be objectified outside the space of reading, thus outside the space of our culture, our time. This involves an emphasis on the textual nature of Bataille's writing and on the intertextual resonances that scan it.

The reading of the book, as object, is problematic, overdetermined by an apparatus which locates and identifies. Without appealing to any space free from such determination by the ideological apparatus of the book's 'circulation', it is necessary to recognize that our reading takes place in a space saturated with meaning, and the text therefore appears as somehow 'at stake' for a culture concerned to mark out its limits and its modes of reading that which may transgress those limits.

/ spilled ink

8. / the influence of anxiety

The anxiety which runs throughout *Histoire de l'œil* and which is signaled in its opening sentence, 'J'ai été élevé très seul et aussi loin que je me rappelle j'étais angoissé par tout ce qui est sexuel'[1] may also be said to characterize its *effect*. The formless inkstain on meaning and communication that the text produces may be likened to the anxiety that characterizes the influence of the text on its readers. As such the 'exscription' which Nancy identifies and which I have qualified as something which dislocates the economy of the text's circulation also describes the more or less undetectable influence of the text on literature and other art forms. The influence of the text is an *influence of anxiety*, to paraphrase Harold Bloom;[2] it is a troubling event in the literature and culture of modernity. Influence here functions not as a benign or Oedipal relation to an authoritative master-text but as something like a trauma, which is repeated.

One might point to writers of the literary erotic for whom Bataille's *Histoire de l'œil* is an implicit precedent: André Pieyre de Mandiargues,[3] Pierre Klossowski,[4] Pauline Réage,[5]

---

[1] *O.C.* I, 13.
[2] Bloom's *The Anxiety of Influence* proposes a theory of influence characterized by an Oedipal tension and its working through.
[3] Cf. André Pieyre de Mandiargues, *L'anglais décrit dans le chateau fermé* (Paris: Gallimard, 1979).
[4] Klossowski's eroticism, in for example novels such as *La Révocation de l'Edit de Nantes* (Paris: Minuit, 1963) and *Roberte ce soir* (Paris: Éditions de Minuit, 1953) would however merit study on its own terms, predominantly, as I suggest below, in relation to Sade.
[5] Cf. Pauline Réage's *Histoire d'O*. The title of Réage's novel is a direct echo of *Histoire de l'œil*, and one could trace a similar determination by the figure of this letter as occurs with the figure of the eye in *Histoire de l'œil* (cf. p. 101). The mediator of the influence of Bataille in this instance is Jean Paulhan, who wrote a postscript for Réage's novel and who was close to Bataille. Réage was the pseudonym of Dominique Aury, Paulhan's lover, the text being originally a kind of love letter addressed to him. Bataille would write a review of the novel for *Critique* in 1955. One might therefore postulate influence here as operating as a kind of communication between men around the text, written

Fernando Arrabal,[6] Bernard Noël,[7] Hervé Guibert,[8] the poetry of Marcelin Pleynet,[9] Yukio Mishima or Angela Carter.[10] But it may be less in terms of aesthetics that we might consider the influence of the text, and rather as the return of certain figures and signifiers. This is certainly identifiable in the visual arts, and primarily in those artists who worked in proximity with Bataille, such as Masson, Miró and Hans Bellmer. For the pre-structuralist writers (Klossowski, de Mandiargues, Réage) it is perhaps the Sadean elements of Bataille's eroticism which are most evident. Indeed, if Sade is an overwhelming, monumental precedent for erotic literature in French, then one might postulate that the influence of Sade in the twentieth century is mediated through Bataille, as a major reader of Sade. The structuralist and post-structuralist reading of Sade, promulgated by *Tel Quel*'s 1967 issue 'La pensée de Sade'[11] (with texts by Barthes, Sollers and Klossowski among others) parallels a different kind of mediation via Bataille, and a different kind of influence. The transgressive text is transgressive now not in terms of what it

by a woman. *Histoire d'O* is an account of the masochistic subjection of its heroine, O, to a series of lovers in a Sadean castle.

[6] A survey of Arrabal's titles suggests the resonance of *Histoire de l'œil* and Bataille's work as a whole: 'Concert dans un œuf', 'La communion solonelle', 'Dieu est-il devenu fou?', 'Bestialité érotique', 'La merde et le ciel'. Arrabal's film *Viva la muerte* features figures which directly echo those of *Histoire de l'œil*.

[7] Cf. Bernard Noël, *Le chateau de cêne* (Paris: Gallimard, 1990).

[8] Guibert's eroticism owes much to Sade, particularly with the texts *Vous m'avez fait former des fantômes* (Paris: Gallimard, 1987) and *Les chiens* (Paris: Minuit, 1982), but certain figures in *Des aveugles* (Paris: Gallimard, 1985) (which features a scene of eye-gouging) and *Le Paradis* (Paris: Gallimard, 1992) (which features a scene of penetration with a revolver) echo those of *Histoire de l'œil*.

[9] Cf. Marcelin Pleynet, *L'amour vénitien* (Paris: Carte Blanche, 1984), *Rime* (Paris: Seuil, 1981). Among the *Tel Quel* writers, Pleynet has signalled his indebtedness to Bataille's fiction rather than to his discursive texts.

[10] Cf. Yukio Mishima, *Thirst for Love* (London: Secker & Warburg, 1970); Angela Carter, *The Bloody Chamber* (London: Vintage, 1995).

[11] *Tel Quel*, 28 (Winter 1967).

represents but in its mode of representation, in its texture. Pierre Guyotat[12] and Bernard Noël,[13] among exponents of the erotic, transgressive text, practise the same kind of displacement of signifiers which *Histoire de l'œil* elaborates. The sexually transgressive text is also a transgression of language; the sexual body is a dismembered, textual body, the scene of pulsional aggression and violence. *Histoire de l'œil* offers the structuralist vision of erotic transgression the precedent of a text which brings together textual and sexual transgression in such a way that they are inseparable. The reading of Sade as a textually transgressive writer, proposed by *Tel Quel* and others, is again mediated through the influence of *Histoire de l'œil*.

I will end this chapter with an example of a plagiaristic citation of *Histoire de l'œil* in another medium which illustrates its seminal importance for proponents of the avant-garde. Jean-Luc Godard's film *Weekend* shows in dark colours the moral corruption and inherent violence of the urban bourgeoisie in 1960s France. Its tone of ludic political subversion resonates with Bataille's violence and the will to 'épater le bourgeois' which characterized the period in which *Histoire de l'œil* was written. But Godard lifts the initial scene of the text, in which Simone sits in the saucer of milk, and sets it in a narrative told, in an indifferent tone, to an inquisitorial male by his mistress, in a darkened room, accompanied and intermittently drowned out by tragic, Wagnerian music, just before they embark on their ill-fated 'weekend'.

Godard's plagiarism removes the scene from the chain of objects which structure Bataille's text, and in doing so repositions the mini-narrative; rather than an author's 'confession' the scene becomes part of a more complex scenario: a woman narrates to her lover her activities with another man and another woman. The scene becomes part of a stereotypical portrayal of the corrupt sexual life of the bourgeoisie, with the

[12] Cf. Pierre Guyotat, *Eden, Eden, Eden* (Paris: Gallimard, 1970).
[13] Noël would edit and preface an edition of the *Documents* articles in 1968.

viewer distanced as addressee from the dialogue. Godard's re-telling of the story plays with the notion of representation and comes out as an essentially tragic, depressive portrayal of human interaction. That Godard reinscribes Bataille's text within his own suggests to what extent *Histoire de l'œil* haunts the transgressive imagination, but, removed from the struc-tural machine of *Histoire de l'œil*, the stakes are different. Godard's film is a discourse upon the social, Bataille's text removed from social space.

The influence of anxiety upon the text's readers is to a large extent a private, unidentifiable matter. But it is evident from the celebration of Bataille as a scandalous, obscene and 'mon-strous' text, in popular culture particularly but also in the discourse of 'theory', that the text of *Histoire de l'œil*, with *Madame Edwarda* and other short texts of Bataille's more transgressive fictions, has provoked an anxiety which has led to a somewhat hysterical response. This hysteria prevents a reading of the text as an analytic, transformative literary oper-ation, and prevents it from being situated in its time. The theoretical analysis and contextual location of this text does not cease to be problematic, however; a counter-argument would criticize the academic enterprise for its desire to appro-priate and locate the text. Bataille has written: 'Il faut le système et il faut l'excès'.[14] It is the tension between the two that characterizes any attempt to write on *Histoire de l'œil*.

9. / the institution

Where does Bataille's writing take place? *Where* does it circu-late? To place Bataille, to situate Bataille is to give him figure. While Bataille writes against figure, towards the *informe*, he

[14] Cited by Sollers in 'Intervention', in Sollers (ed.), *Bataille*, 10.

writes from *somewhere*. And specifically: from the institution, from within the mural, whose boundaries are outlined on a map. The institution is a privileged figure in Bataille's writings. Bataille is the librarian, inside the Bibliothèque Nationale. To put Bataille on the map, then. But in so doing, to show that the place, the *where*, turns into a series of places, which are in relation to each other, a network. The mural becomes not the enclosed, the outline of a figure which is walled in, but the labyrinth, another name for the network. Turn the map into a labyrinth, a network. The movement to the *informe*, the breakdown of figure, happens underneath the institution, in the labyrinth, *between* the various sites of Bataille's activity. Or rather, not underneath, in the same place, on the same ground, but with a completely different conception of space, relational and labyrinthine. The objective is to identify a series of writings, locate them, establish the theory of a different conception of space, establish a network both geographic and historical, cultural, for Bataille's activities around *Histoire de l'œil*.

Georges Bataille is born in 1897. The family will move to Reims, where Bataille attends a lycée until 1913. He spends the year 1917–1918 in a seminary. In November 1919 he applies and is admitted to the Ecole Nationale des Chartes in Paris, where he studies for the next three years, submitting a thesis titled *L'ordre de la chevalerie, conte en vers du XIIIème siècle avec introduction et notes*, in 1922. After the success of his thesis he is named as an 'archiviste-paléographe'. In February 1922 he studies at the Ecole des Hautes Etudes Hispaniques in Madrid. In June 1922 he is appointed librarian at the Bibliothèque Nationale and takes up the post the next month. In 1924 he is appointed as librarian in the Department of Coins and Medals.

Bataille's extant writings up to this point are varied and heterogeneous. They begin with the rediscovered text *Notre Dame de Rheims*, written and published in 1918.[1] A projected

---

[1] Included in Hollier, *La prise de la Concorde* (Paris: Gallimard, 1974).

collaboration with the philosopher Leon Chestov is abandoned, but he will be involved in a translation of Chestov's *L'idée du bien chez Tolstoï et Nietzsche*. In 1926 he starts to collaborate on the art and archaeology review *Aréthuse*, and will contribute a number of short articles and reviews to it. A series of 'Fatrasies', prepared by him but without his name, appear in *La révolution surréaliste* in 1926, Bataille's friend Michel Leiris having acted as intermediary between Bataille and André Breton. The manuscript of *W.C.*, written in 1926, is destroyed, but some fragments from it will appear as an 'introduction' to *Le bleu du ciel*, under the title *Dirty*. In January 1927 he writes the text *L'anus solaire*.

There is an immediate impression of heterogeneity. The obscene writings *L'anus solaire* and *Histoire de l'œil* exist alongside the institutional activities as an archivist, paleographer (working on manuscripts) and numismatist at the Ecole des Chartes and subsequently at the Bibliothèque Nationale. Bataille's thesis, *L'ordre de la chevalerie*, and the notes published in *Aréthuse* are scholarly, communications within the discipline. They are marked by the language of museography and scientific classification. Denis Hollier, in *La prise de la Concorde*, remarks: 'Sans doute n'y a-t-il pas lieu d'attacher une importance excessive à ces articles dont il semble bien qu'un zèle professionelle soit seul à l'origine.'[2] But what determines the passage from 'professional zeal' to a different, scatological and violent, kind of zeal?

A series of key dates in Bataille's early writings can be established:

1918: Writes *Notre Dame de Rheims*
1922: Writes *L'ordre de la chevalerie*
1925: Starts psychoanalysis with Dr Adrien Borel
1926: Submits 'Fatrasies' to *La révolution surréaliste*
       Writes *W.C.*
       First contributions to *Aréthuse*

[2] Hollier, *La prise de la Concorde*, 228.

1927:	Further contributions to *Aréthuse*
	Writes *L'anus solaire*
	Further contributions to *Aréthuse*
	Begins *Histoire de l'œil*
1928:	Further contributions to *Aréthuse*
	Publication of *Histoire de l'œil*
1929:	Writes *L'Amérique disparue*
	Final contribution to *Aréthuse*
	Documents no. 1: *Le cheval académique*

Entries into this series can establish relations between them, drawing out a web of interrelated figures and points of intersection, establishing a network.

First entry into this series

Bataille's thesis, *L'ordre de la chevalerie*, is a presentation, introduction and annotation of a versified *conte* from the thirteenth century. It is essentially concerned with an 'explication symbolique des rites de l'adoubement et de l'enseignement des devoirs des chevaliers',[3] in which 'la chevalerie est regardée telle qu'un saint ordre reçu par le chevalier.'[4] *Adoubement*: the reception of a sacred armour. Bataille remarks in the thesis how the sacred meaning attached to the gift and the wearing of armour derive from an interpretation of a text of Saint Paul, 'où celui-ci incite les chrétiens à se revêtir de l'armure des vertus chrétiens'.[5] In 1939, Leiris will publish *L'âge d'homme*, which he characterizes as the passage into 'l'ordre féroce de la virilité',[6] dedicated to Bataille.[7] Certainly, Bataille's involvement in and creation of secret societies echoes the order of

---

[3] *O.C.* I, 99.
[4] Ibid., 101.
[5] Ibid., 100.
[6] Michel Leiris, *L'âge d'homme* (Paris: Gallimard, 1939), 42.
[7] Bataille had suggested to Leiris that he write an erotic version of his autobiography to appear in a collection of erotic literature. Although the project was never completed, *L'âge d'homme* is conditioned by this demand. Cf. Leiris, *Journal* (Paris: Gallimard, 1992), 875.

knighthood, and these are studied in the contributions to the *Collège de Sociologie* (the numismatics articles in *Aréthuse* are also concerned with secret societies). The entry into a secret order begins with a sacred naming (*adoubement*: dubbing), and the gift of a sacred armour. *Histoire de l'œil* will offer a profane mirroring of this rite: the stripping of armour, the *mise à nu* will become sacred; 'Sir' Edmond 'blesses' the priest Don Aminado with the profaned host. Simone and Marcelle receive the body's waste (sperm, urine) as a sacrament. Between the two texts a reversal operates, from pure to impure sacred. *Histoire de l'œil transforms* the previous writings.

Second entry into the series

If we look at the articles written for *Aréthuse* up to 1928, although some of their concerns will become important in Bataille's later writings (the question of sovereignty, the representation of the sun as divine force, the association of the sovereign with animality), the difference between these notes and the articles which would appear in *Documents*, starting from 1929, is marked. The *Aréthuse* articles concern the index of the coin as an aid to the historical and political knowledge of the periods in question. They are concerned with sovereignty as legislative power; the ruler in control of the coining is the legislative ruler. In Bataille's later writings, and particularly in the *Collège de Sociologie*, by which time Bataille would have been familiar with Dumézil's work on the different representations of sovereignty in early Indo-European culture (Mitra/Varuna),[8] the sovereign takes on another role: that of incarnating excess, the sacred, in his (or her) body: the sacrifice or death of that body giving rise to manifestations of sexual and other kinds of transgression. In other words, if the *Aréthuse* articles and the thesis essentially deal with *le sacré faste*, as legislative and political embodiment of the sacred, the sub-

[8] Cf. Georges Dumézil, *Mitra-Varuna* (Paris: Gallimard, 1948), and Hollier (ed.), *Le Collège de Sociologie* (Paris: Gallimard, 1995).

sequent writings foreground *le sacré néfaste*, the carnal, 'low' elements of the sacred. A key text is *L'Amérique disparue*,[9] roughly contemporary with the publication of *Histoire de l'œil* and preceding the *Documents* texts, published in the *Cahiers de la République des Lettres*, specifically in a special issue devoted to an exhibition of pre-Columbian art. Its first paragraph is remarkable for its verbal violence:

> La vie des peuples civilisés de l'Amérique avant Christophe Colomb n'est pas seulement prodigieuse pour nous du fait de leur découverte et de leur disparition instantanées, mais aussi parce que jamais sans doute plus sanglante excentricité n'a été conçue par la démence humaine: crimes continuels commis en plein soleil pour la seule satisfaction de cauchemars déifiés, phantasmes terrifiants! Des repas cannibales des prêtres, des cérémonies à cadavres et à ruisseaux de sang, plus qu'une aventure historique évoquent les aveuglantes débauches décrites par l'illustre marquis de Sade.[10]

This difference in tone from the *Aréthuse* contributions, which will be continued in the *Documents* articles, is not due simply to a difference in the place of publication, to a movement outside the institution. The *Cahier de la Republique des Lettres* was run by virtually the same editorial team as *Aréthuse*, as was *Documents*, although one might suppose Bataille's greater involvement in the inception of *Documents* afforded him greater freedom. His disputes with Pierre d'Espezel (co-founder of *Documents*, director of the *Cahier* and co-director, with Jean Babelon, of *Aréthuse*) show however that the more violent and scatological tone of writings subsequent to 1928 is not taken lightly by the scholars, and that the shift to the different register is part of a development in Bataille's thinking and writing.

[9] *O.C.* I, 152–8.
[10] Ibid., 158.

Third entry into the series

As Denis Hollier shows in his reading of *Notre Dame de Rheims*,[11] what its author is concerned to do is to show the cathedral as the incarnation of the history and geography of France, that is, as the incarnation of political, legislative power, that of the Nation, bathing also in the sacred light of God. The cathedral, threatened by the heterogeneous, violent force of war, remains the incarnation of the Nation: 'Disloquée, vide et défigurée, la Cathédrale est toujours la France.'[12] For Deleuze and Guattari, reading Dumézil, war is 'bien irréductible à l'appareil d'Etat, extérieure à sa souveraineté, préalable à son droit, elle vient d'ailleurs . . . Il serait plutôt comme la multiplicité pure et sans mesure, la meute, l'irruption de l'éphemère et puissance de la métamorphose.'[13] More than the Oedipal opposition between the maternal body and the *nom du père*, 'synonyme de la guerre',[14] what Bataille opposes here is the legislative power of the Nation and the violent, non-capitalized force of war, 'inexorable comme est la guerre',[15] and this opposition corresponds to the two aspects of the sacred, the incarnated and legislatively just form which has received God's blessing (*adoubement*), and the force of pure multiplicity and metamorphosis, *rhizome*: violence as sacred, as represented in the opening to *L'Amérique disparue* cited above. Against the reading of the cathedral or of coins as an index to history, history as the catalogue of legislative power, pre-Columbian art, for example, bears witness to rituals which are 'plus qu'une aventure historique'.[16] If the force of the word alone is anything to go by, the vocabulary associated with the (re)introduction of the sacred as force of metamorphosis is the following: *démence, crime, soleil, cauchemar, phantasme,*

---

[11] Hollier, *La prise de la Concorde*, 45–52.

[12] Ibid., 41.

[13] Gilles Deleuze and Felix Guattari, *Mille plateaux* (Paris: Minuit, 1980), 435.

[14] Hollier, *La prise de la Concorde*, 49.

[15] Bataille, *Notre Dame de Rheims*, in Hollier, *La prise de la Concorde*, 39.

[16] *O.C.* I, 152.

terreur, cannibalisme, cadavre, prêtre, sang, débauche, aveuglement, Sade.

Fourth entry into the series

In the first note Bataille wrote for *Aréthuse*, a review of a book by Charles Florange on 'messageries et postes d'après des documents métalliques et imprimés',[17] he notes how the 'documents' highlight 'l'effort accompli autrefois pour organiser un magnifique réseau de circulation'.[18] *Documents* will privilege use-value over 'circulation', exchange, as Denis Hollier describes in a key article,[19] and as the title of a later article, 'La valeur d'usage de D. A. F. de Sade' suggests.

Fifth entry into the series

The *Aréthuse* articles focus on animals as numismatic indexes or symbols of the sovereign's power, or of the zodiac. The articles in *Documents*, and *L'Amérique disparue*, offer a vision of a *becoming animal*, according to Deleuze and Guattari's term: 'Ce qui est réel, c'est le devenir lui-même, le bloc du devenir, et non pas des termes supposés fixes dans lequel passerait celui qui devient. Le devenir peut et doit être qualifié comme devenir-animal sans avoir un terme qui serait l'animal devenu.'[20] The evolution of the representation of the animal form is proposed as analogous to the evolution of civilization, in such a way that the passage from the primitive form of the horse to its 'academic' form mirrors that from monkey to man. Bataille: 'Et il importe d'observer à ce sujet que les paléontologues admettent que le cheval actuel dérive de lourds pachydermes, dérivation qui peut être rapprochée de celle de l'homme par rapport au hideux singe anthropomorphe'.[21]

[17] Ibid., 107.
[18] Ibid.
[19] Hollier, 'The use-value of the Impossible', in *October*, 60 (Spring 1992).
[20] Deleuze and Guattari, *Mille plateaux*, 291.
[21] *O.C.* I, 162.

History appears now as not so much a process of civilization, of evolution from lower forms to higher forms, but an oscillation between the two: 'tantôt l'effroi de ce qui est informe et indécis aboutissant aux précisions de l'animal humain ou du cheval; tantôt, dans un tumulte profond, les formes les plus baroques et les plus écœurantes se succédant.'[22] A dynamic of oscillation between the human and the animal which echoes the swing of Giacometti's suspended ball, theorized by Krauss as the movement of the *informe*.

Sixth entry into the series

'Vous avez ouï-dire de Rheims qui fut une grande ville dans la plaine de Champagne. Elle avait une histoire antique: Clovis barbare, que baptisa Saint-Rémy donnait à la bonne ville chrétienne un pieux prestige et l'on y sacrait les rois de la France.'[23] Clovis, baptized in Reims in 496, was the first king of the Franks. The first sentence of Bataille's first published text *Notre Dame de Rheims*, the first sentence written by Bataille, as Hollier remarks,[24] recalls the inauguration of the nation of France, governed by the Franks. 'Le cheval académique', published in the first issue of *Documents* eleven years later, is a study of coins produced by the Gauls which represent horses. Here it is the Gauls, later invaded and overcome by the Franks, whose culture is affirmed by Bataille: 'Il est facile d'opposer aux conquêtes systématiques des Grecs ou des Romains, les incursions incohérentes et inutiles des Gaulois à travers l'Italie ou la Grèce et, en général, à une constant faculté d'organisation, l'instabilité et l'excitation sans issue.'[25] Pre-1928, Bataille's writing inaugurates itself with a hymn to the sacred inauguration of the kings of France, their *adoubement*; post-1928, his writing lends its voice to the incoherent, *informe*

---

[22] Ibid., 163.
[23] Bataille, *Notre Dame de Rheims* in Hollier, *La prise de la Concorde*, 35.
[24] Ibid., 45.
[25] *O.C.* I, 160.

extravagance of the Gauls. In the debate about French racial identity with which Rimbaud engages in 'Mauvais sang',[26] Bataille changes sides. This has further consequences: if Bataille's writings begin with the inauguration of the nation state, France, then subsequent to 1928, with 'Le cheval académique' he writes of the Gauls as a force of random violence: 'Tout ce qui peut donner à des hommes disciplinés une conscience de valeur et d'autorité officielle: architecture, droit théorique, science laïque et littérature de gens de lettres, était resté ignorés des Gaulois qui ne calculaient rien, ne concevant aucun progrès et laissant libre cours aux suggestions immédiates et à tout sentiment violent.'[27] The Gauls are a Deleuzian war machine: 'une indiscipline fondamentale du guerrier, une remise en question de l'hiérarchie, un chantage perpétuel à l'abondon et à la trahison, un sens de l'honneur très susceptible et qui contrarie, encore une fois, la formation d'Etat . . . forme pure d'extériorité.'[28] Instead of inaugurating the state, threatened (again) by the Germans (the Franks), Bataille sets out the opposition of the classical horse as formal perfection of the organized state and the 'ignobles singes et gorilles équidés des Gaulois'[29] as the *informe* representation of the war machine. Henceforth the noble and classical contours of form will appear as a line of neurotic defence against the violence of 'le bas', a territorialization (an outline) closed up against the contagious flight-lines of deterritorialization.

1927–8 appears a crucial moment in Bataille's trajectory as a writer. Although he has explored *le sacré néfaste* in a fictional text, in *W.C.*, this is an unidentifiable, unpublished and fragmentary text. The writings published under the name

[26] Arthur Rimbaud, 'Mauvais sang', in *Une saison en enfer* in *Poésies* (Paris: Livre de poche, 1963), 108: 'J'ai de mes ancêtres gaulois l'œil bleu blanc, la cervelle étroite et la maladresse dans la lutte. [. . .] Les Gaulois étaient les écorcheurs de bêtes, les brûleurs d'herbes les plus ineptes de leur temps.[. . .] Si j'avais des antécedents à un point quelconque de l'histoire de la France! Mais non, rien. Il m'est bien évident que j'ai toujours été de race inferieure.'
[27] *O.C.* I, 160.
[28] Deleuze and Guattari, *Mille plateaux*, 443.
[29] *O.C.* I, 162.

'Bataille', *Notre Dame de Rheims*, *L'ordre de la chevalerie*, and the *Aréthuse* articles privilege the essentially Christian, sacred order of *le sacré faste*. The hinge between the writings of the pure sacred and the impure sacred, between state power and the Deleuzian war machine, or, to use our initial figure, between the mural and the labyrinth, finds its location in the years 1927–8: a time which sees Bataille's analysis with Adrien Borel, Borel's transmission to him of the photos of the *Supplice des cents morceaux* (of which more later), the writing of *L'anus solaire* and *Histoire de l'œil*. It is around this nexus that the transformation occurs.

10. / centrifugal / centripetal

Before mapping out Bataille's movements in geographic and cultural space, I propose to analyse the text of *L'anus solaire*, written in 1927, as a theorization of space and movement which is the blueprint for *Histoire de l'œil*, the matrix which it will 'apply'. *L'anus solaire* is cited by critics as one of the texts which led Dr Dausse, Bataille's doctor, to propose that his patient was 'sick' and that he should seek a psychoanalytic cure with Dr Adrien Borel. But what are the basic elements of this 'pathology'? It proposes *parody* as fundamental:

> Il est clair que le monde est purement parodique, c'est à dire que chaque chose qu'on regarde est la parodie d'une autre, ou encore la même chose sous une forme décevante.[1]

Things are parodies of other things. How is it possible to understand this? Fetishism? the object is a substitute for another thing. Metaphor? the object stands for another thing. Metonymy? the object is a displaced element of another thing. The *Petit Robert* offers: 'contrefaçon grotesque, carica-

---

[1] *O.C.* I, 81.

ture, travestissement'. Objects and words appear as isolated from one another, contained within their contours, but a deeper level of continuity exists where each object is linked to another, in a contagious movement of displacement. This movement is *parodic*, that is, a travesty, a transgression. The movement of contagion between objects or words necessarily breaks the outline of form, the contours, and in doing so it parodies, travesties that form (the movement of relation breaks through the walls of the mural). Objects, words, are in a constant state of leakage ... This is not the Platonic model of a vertical relation of a form to an idea, which is why the models of fetishism and metaphor do not work here, but a horizontal movement. There is in every object a leakage, a contagiousness which empties it out in a movement towards another object, or indeed towards a travestied form of itself. Language and thought are thus characterized by a relationality, a network, or in spatial terms a labyrinth. There is also the utopian notion of a unique glance to which would be revealed the network in its totality:

> Depuis que les phrases *circulent* dans les cerveaux occupés à reflechir, il a été procédé à une identification totale, puisque à l'aide d'un *copule* chaque phrase relie une chose à l'autre; et tout serait visiblement lié si l'on découvrait d'un seul regard dans sa totalité le tracé laissé par un fil d'Ariane, conduisant la pensée dans son propre labyrinthe.[2]

From this we can pick out a series of key terms in order to reformulate an aesthetic vision: *circulation, copula, labyrinthe, regard*. Sentences circulate, via the copula, making thought a labyrinth. The utopian horizon of Bataille's vision is a totality, a circle, where, in a single glance, the totality of relations would be revealed. The round forms of *Histoire de l'œil* are microcosmic parodies of this macrocosmic grand circle, the total circle of relations. But their closure is continually broken, copulated; *circulation* does not *yet* form the total circle of relations. Thought is not a circular tower, or a pyramid, but a

[2] Ibid.

labyrinth. Two principles: the total circle of relations and the labyrinth of thought in copulation. But the grand circle of the copulation of all objects, the circulation of all sentences, is hinted at in a single glance. The rupture of closed circularity on a microcosmic scale, the internal collapse of the egg, for example, reveals to the eye, for a moment, the terrible possibility of the totality of relations, without limits. The anxiety caused by this look, the look of the protagonists of *Histoire de l'œil*, is an anxiety at the possibility of the lack of any limits, of any closure, or any possibility of isolation, thus of any possibility of relation. The terror that the world is *informe*, pure movement, pure flux without stasis.

Language, then, *les phrases*, is the *fil d'Ariane* which weaves out this network of relations between objects. Bataille's proto-structuralist proposition, similar to Derrida's *différance*, is that the relational nature of language determines a contagious, parodic view of the world. Language operates through the linkage/leakage of the *copula*, through copulation, both acts forcing a rupture of the limits of form. Bataille's view of language as copulative network is mirrored and determined by the picture of sex in *Histoire de l'œil*. Language is fundamentally linked to sex, not only through mimesis, but through the body:

> Mais le copule des termes n'est pas moins irritant que celui des corps. Et quand je m'écrie: JE SUIS LE SOLEIL, il en résulte une érection intégrale, car le verbe être est le véhicule de la frénesie amoureuse.[3]

Sex is not predominant. The copulation of bodies is, like language, a relation of unity and closure to fluidity, a movement out of limits and a movement back into them. Like language, sex enables that glance, *le seul regard*, which would incite the anxiety of pure fluidity. Pure *jouissance*.

Bataille's cosmic vision: you can break into this circulation at any point. The centre is not fixed. The circle has no centre.

---

[3] Ibid.

There is no gold standard. The earth is not the centre. The earth moves around the sun. A rotational movement. At the same time, a copulative movement:

> Et si l'origine n'est pas semblable au sol de la planète paraissant être la base, mais au movement circulaire que la planète décrit autour d'un centre mobile, une voiture, une horloge ou une machine à coudre peuvent également être acceptées en tant que principe générateur.[4]

Two principles of movement: rotational and copulative. They change into one another constantly. Their combination is figured by a locomotive composed of wheels (circulation) and pistons (copulation).

Parody:

> Un soulier abandonné, une dent gâtée, un nez trop court, le cuisinier crachant dans la nourriture de ses maîtres sont à l'amour de que le pavillon est à la nationalité.
> Un parapluie, une séxaguinaire, un séminairiste, l'odeur des œufs pourris, les yeux crevés des juges sont les racines par lesquelles l'amour se nourrit.[5]

Every object is determined by two axes: the rotational axis, the grand cosmic circle, and the copulational axis. While the latter is determined by forgetfulness, the umbrella forgets itself in its movement towards the sewing machine, the egg is forgotten by the eye, in the long term all things join up on the utopian horizon of a cosmic vision.

> Le mouvement est la figure de l'amour incapable de s'arrêter sur un être en particulier et passant rapidement de l'un à l'autre.
> Mais l'oubli qui le conditionne ainsi n'est qu'un subterfuge de la mémoire.[6]

The earth revolves around the sun. The sun is the principle of rotational movement. But on earth the movement of copulation means a state of constant forgetfulness, the collapse of one

---

[4] Ibid., 81–2.
[5] Ibid., 82.
[6] Ibid., 83.

thing into another, its death. Rotational movement and copulative movement interact in such a way that the object is determined on two axes, a vertical axis of erection which links the object to the sun, and a horizontal axis of copulation which links one thing to another, *branlement*. Bataille's images are generated by a desire to show the intersection of both axes. Thus the anus is the sun. The erection is a tree which will fall. The volcano is an anus ejecting waste from the bowels of the earth. The penis entering the vagina is the figure, in copulative movement, of the rotational movement of the earth around the sun.

Thus men turn their eyes away from the sun, corpses, coitus. Men's eyes are not turned towards the sun: they are on the horizontal axis of copulation. The sun is also a night because to look at it is to induce blindness through intense luminosity. Coitus is the figure of the sun. The anus is the night which is a sun. *Histoire de l'œil* fixes the eye on these scenes, deliberately and systematically.

Bataille's aesthetic theory is initially proposed in cosmic terms. His aesthetic theory is indissociable from a theory of the cosmos characterized by a Heraclitean flux (among Bataille's loans from the BN are several volumes on pre-Socratic thought and particularly on Heraclitus),[7] but also by copulative movement. It is not so much a philosophy as a theory of movement and a vision of thought determined by an expansion of the sexual: sex becomes the figure of the forces at work in the cosmos. But this cosmic theory entails a certain view of language (as *fil d'Ariane*), and a thematics of vision, which *Histoire de l'œil* will engage.

Bataille's vision is highly singular and no doubt determined by his biography, by certain events, and, in psychoanalytical terms, by a certain pathology, which I will consider in the light of an analysis of the second part of *Histoire de l'œil*, the 'autobiographical' 'Coïncidences'. It is also conditioned, as

[7] Cf. *O.C.* XII.

proposed above, by an inscription in and against the institution, by a transformation of institutional writings (writings
which institute, which incarnate), their disincarnation. But
while Bataille comes from *outside* the literary and artistic
milieux of Paris (from *inside* the institution), in the mid-
twenties he enters into these milieux, and, if he marks them
more than they mark him, the aesthetic concerns of these
milieux also condition Bataille's aesthetic vision and literary
style. This is the subject of the subsequent section.

11. / a map

The map of cultural space in 1920s Paris features certain
'zones of intensity', to borrow again the terminology of Deleuze
and Guattari's *Mille plateaux*. Bataille comes from within the
institution, specifically the Bibliothèque Nationale Department of Coins and Medals, but the non-institutional milieux
in which *Histoire de l'œil* will circulate are sites of Surrealist
dissidence. In mapping out Bataille's initial intervention into
the field of the 'literary', these sites are crucial. But to be
added to them are the institutional sites of Bataille's first
writings and certain extrinsic locations which are specific to
Bataille's biography; they form a kind of network which is
Histoire de l'œil's most immediate context.

    1. The site of the institution: Bibliothèque Nationale, *rive
droite*, Rue de Richelieu. As set out above, the Bibliothèque
Nationale is the realm of numismatics, the institutional review
Aréthuse (edited by Bataille's colleagues there Pierre d'Espezel
and Jean Babelon). It is mirrored by an extrinsic site, the
British Museum in London, which Bataille will visit for purposes of research (the *Aréthuse* articles include considerations
of the BM's collections). A colleague at the Bibliothèque
Nationale, Jacques Lavaud, will introduce him to Michel Leiris

in 1924, who will introduce him to André Masson and the group around 45 Rue Blomet (see **5**).

2. Another institution: the Ecole des Chartes, 19 Rue de la Sorbonne, *rive gauche*: where Bataille completed his thesis and from which he entered into the BN. A colleague there, Alfred Métraux, will become a friend, a contributor to *Documents* and a later acquaintance of Leiris in the context of anthropology (see **3**).

3. Another institution: the Musée de l'Homme, *rive droite*, opposite the Tour Eiffel. A number of individuals associated with the Musée de l'Homme will become important contributors to *Documents*: André Schaeffner, Marcel Griaule, and eventually Michel Leiris. Its deputy director, Georges Rivière, will also be one of the founders of the review.

4. The pole of the Surrealist movement: Rue Fontaine, Place Blanche, *rive droite*, André Breton's 'magnetic personality'. In 1925 Michel Leiris and André Masson will join this group, through the intermediary of Georges Limbour (see **5**), only to be excluded from it in 1929 when Breton attempts to cement and circumscribe the commitment of the Surrealist group to collective action (at the 'Bar du Chateau', see **6**).

5. Site of Surrealist dissidence: 45 Rue Blomet, *rive gauche*. André Masson lived there alongside Joan Miró. They had knocked a hole in the wall to allow for easier communication between their two apartments. The building becomes a meeting place for a loosely defined group of which the most constant members are Leiris, Roland Tual, Georges Limbour and André Masson. Others who pass through it include Artaud, Gertrude Stein, Hemingway, Roussel. Bataille frequents this site from 1924, introduced to Masson by Leiris, and to Leiris by Lavaud (see **1**), and will continue to do so despite Leiris, Masson, Tual, Limbour and Artaud's adherence to the Surrealist group, centred around the Rue Fontaine (see **4**).

6. Second site of Surrealist dissidence: 54 Rue du Chateau, *rive gauche*. A Surrealist 'annexe' which at various times has a number of different members of the group living there: Marcel Duhamel, Yves Tanguy, Jacques Prévert initially, then Louis Aragon, Robert Desnos, André Thirion and others. Bataille will frequent 54 Rue du Chateau in 1926 and 1927. Many of the members of the Surrealist group who were associated with the Rue du Chateau and the Rue Blomet groups will become associated, after 1929, with *Documents*, and will be signatories of the violent tract directed against Breton, *Un cadavre*.

7. The *Bureau des recherches surréalistes*: Rue de Grenelle, *rive gauche*, set up in 1924, run by Gérard and subsequently Artaud. In 1928 the Bureau hosts discussions on sexuality, in which it becomes evident that Breton's conception of sex is highly moral. To the question, 'What about animals?' 'You're joking', total refusal and disgust at homosexuality (against Queneau), 'I refuse to have recourse to artificial methods when it's a question of love, and it is a moral issue for me. The alternative would be libertinism'. Refusal of the presence of a third person. Breton's high moral tone contrasts with the perverse interests of Surrealist dissidents, and the entire approach of Bataille.

8. An extrinsic site, in Spain: the *arena* in Madrid, where on 7 May 1922 Bataille will witness the enucleation of the bullfighter Granero.

9. A further extrinsic site: Siena cathedral, which Bataille visited in 1923, his fragmentary writings on it marking his loss of faith.

10. A brothel in Saint Denis, where Leiris recalls that Lavaud, Bataille and himself would plan a review titled *Oui*, countering Dada's 'Non', with contributions by the girls.

11. An extrinsic site, in London: the Savoy Hotel, where the introductory episode of *Le bleu du ciel*, believed to be a surviving fragment of *W.C.*, from 1926, is set.

12. *Nord-Sud*. Between the right bank and the left bank runs the Clignancourt-Orléans Metro line, called the *Nord-Sud* line, an axis of communication between the magnetic north of Surrealism and the dissident left bank. *Nord-Sud* is also the name of a review run by Pierre Reverdy between 1917 and 1918, publishing Apollinaire, Max Jacob and early texts by the future Surrealists, which forms the immediate pre-history of the avant-garde in the period we are concerned with.

The cultural context in which *Histoire de l'œil* appears is historically 'covered' by the movement of Surrealism. André Breton's *Nadja* appears in the same year (1928) as Louis Aragon's anonymous erotic text *Le con d'Irène*, like *Histoire de l'œil* published by René Bonnel and illustrated by André Masson. Dalí and Buñuel's film *Un chien andalou* appeared early in 1929. Breton's *Manifeste du Surréalisme* had appeared two years previously, and many of Bataille's associates (Masson, Leiris, Tual) had rallied to the Surrealist movement. The history of the dissidence from the movement is well documented,[1] and the role of the review *Documents*, in which Bataille was to play a prominent part, as a forum for an explicitly materialist, desublimatory critique of Surrealism has been celebrated.[2] Many of the writers and artists who would either be excluded or exclude themselves from the Surrealist movement, and whom Breton would attack in the second manifesto, would write for *Documents*, including Michel Leiris, Robert Desnos and Jacques Prévert. Bataille himself, despite having contributed, at Leiris's request, some edited medieval 'Fatrasies' to *La révolution Surréaliste*, would always remain distant from and critical of the Surrealist movement and specifically Breton.

The tendency to characterize the whole period as 'surrealist' in a generalized manner acts to obscure the specificity and punctuality of the complex network of relations that make up this context, or these, plural, contexts. In effect, even 'Surrealism' as a movement is characterized by a plurality of different groups and milieux, and this plurality is perhaps what conditions the exclusions, departures and polemics that characterize the movement. The plurality of the movement

[1] Cf. André Masson, *Le rebelle du Surréalisme: Ecrits* (Paris: Hermann, 1976); M. Surya, *Georges Bataille, la mort à l'œuvre* (Paris: Gallimard, 1992); Jean-Louis Houdebine, 'L'ennemi du dedans', in Sollers (ed.), *Bataille*.
[2] Cf. Krauss, *The Optical Unconscious*, and her earlier book, *The Originality of the Avant-Garde and Other Myths* (Cambridge, Mass: MIT Press, 1985).

can be precisely mapped out in terms of location, as in the map above, for if the centralized pole of the Surrealist movement, the site of its identity and its cohesion, is the Cyrano Café, Rue Fontaine, where Breton would organize meetings, this centre is offset by other locations, other fora of activity, on the left bank, which have a much more fluid, less circumscribed and identified nature. Two such locations, 45 Rue Blomet and 54 Rue du Chateau, quite close to one another, play a part in Bataille's *entrée en matière*, or at least his entry into the artistic and literary milieu of 1920s Paris.

The first of these, 45 Rue Blomet, is particularly important, as it forms an alternative nucleus, almost a second movement in the shadow of Surrealism, but not seeking to define itself according to any principles, particularly scientific or pseudo-scientific ones. Masson will refer to it as *'le mouvement flou'*,[3] and this 'vagueness' can be taken as a positive affirmation of the fluidity that characterized the activity of the Rue Blomet and as a description of its aesthetic.[4]

While, as already mentioned, many of the associates of the Rue Blomet would adhere to the Surrealist movement only to leave it a few years later, in 1929 (the contact was made unwillingly by Georges Limbour, who frequented both groups), a difference in approach is clearly marked out. In his *Conversations avec Georges Charbonnier*, and in many *compte-rendus* of this time, Masson recalls that of the trinity of names celebrated at Rue Blomet, Sade, Nietzsche and Dostoyevksy, only the first would meet with Breton's sympathy, Nietzsche and Dostoyevksy provoking a violent antipathy.[5] The Rue Blomet is more resonant with Bataille's concerns; his interest in Nietzsche and Dostoyevsky, provoked by his work with Chestov, would have found a response there. Leiris describes the opening of *Le bleu du ciel*, the early text Bataille is reputed to have written in 1926, as very much in the style of Dostoy-

---

[3] Masson, *Conversations avec Georges Charbonnier* (Paris: Julliard, 1958).
[4] Ibid.
[5] Masson, *Ecrits*, 77.

evksy's *Le sous-sol*.[6] The taste at Rue Blomet for reciting Elizabethan tragedies is resonant with Bataille's tragic laughter. Rue Blomet affirmed humour, a perverse eroticism, and elements of a modernity that was beginning to make itself felt in Paris at this time: jazz, *l'art nègre* (the *Bal nègre* was next door), while Rue Fontaine celebrated moral rigour in its revolt, *l'amour fou* (the discussions of the *Bureau des recherches surréalistes* on sexuality, as suggested above, show very clearly Breton's distaste for sexual perversity and his adherence to an ideal of pure love). If it is Masson who makes the first experiments with automatism in his drawings, automatism, in the Rue Fontaine and in the Surrealist publications, is theorized and justified according to the science of psychoanalysis, albeit partially or erroneously applied.

Bataille's *Histoire de l'œil* should thus be dissociated from a generalized 'surrealism' and read in relation to the artistic and literary milieu from which it emerges: Rue Blomet. This is not to say that the Rue Blomet is the sole context which determines the text, or indeed that the context is determinative of the text in a linear fashion. Bataille is specifically not part of the artistic milieu or context of any particular site, Leiris's circle of friends, partially originating from fairly well-off family and college friends, is specifically not the milieu from which Bataille comes; he comes from *somewhere else*, as we have seen: it is the intersection of the concerns specific to Bataille and the cultural context of the Rue Blomet which inform *Histoire de l'œil*. To the extent that the text is influenced by a milieu into which Bataille is introduced and which he will frequent, and which he will influence, a milieu with certain terms of reference and a precise aesthetic, *Histoire de l'œil* is a product of this context, that is to say, a reading of this aesthetic context.

This implies a separation of the aesthetics of the Rue Blomet from those of 'Surrealism', and an identification of the

---

[6] Michel Leiris, 'De Bataille l'impossible à l'impossible *Documents*', in *Critique*, (Aout-Septembre 1963) 195–6, 687.

terms of reference and the punctual set of relations which defines it, without losing sight of those aspects of the period which were common—a desire for modernity, a desire to shock and for revolt, a distaste for 'literature'. As suggested above, the distinction may be expressed in terms of a different conception of space and movement. While Breton and the 'official' Surrealist nucleus was centrifugal, centralized around a 'magnetic pole' (Breton), Rue Blomet was dispersed, non-centralized, non-hierarchical. While the reaction of the former to the notion of structural or unconscious determination is the more or less rigid definition of 'automatic writing' and of its appearance as a 'champ magnétique', for the latter this determination is not defined or magnetized but ludic and proliferating, parodic. In *Histoire de l'œil* and in Masson's drawings, or Miró's, in Leiris's *Aurora*,[7] movement is centripetal, contagious. The map does not simply show the right bank as opposite the left bank; they are entirely different organizations of space. Rue Fontaine is the magnetic pole of a series of concentric circles. The left bank is a Heraclitean space of flux, *tourbillons*, Brownian motion. Mapping out their difference is a question of considering different conceptions of topographical movement, different conceptions of relation.

For example, if *Nadja* represents the topography of Paris as a series of polarized sites where the narrator meets his repressed past in the form of an uncanny return,[8] the topography implied by the writings produced by Leiris, Limbour and Bataille is not situated, it is a topography of displacement and effraction, passage from one site to another; in *Histoire de l'œil*, the protagonists are always on the move, and the door, the entry into and the movement out of closure becomes the privileged site of action. Space is not the magnetic

[7] Republished as Michel Leiris, *Aurora* (Paris: Gallimard, 1973).
[8] Cf. Margaret Cohen, *Profane Illumination: Walter Benjamin and the Paris of Surrealist revolution* (Berkeley and Los Angeles: University of California Press, 1993) and Patrick ffrench, 'Dérive: the *détournement* of the *flâneur*', in *The Hacienda Must be Built: On the legacy of situationist revolt* (Manchester: Aura, 1996).

pole which attracts the subject back to a repressed or forgotten past, but a fluid field of investment and disinvestment. *Histoire de l'œil* is concerned with spaces to be displaced, exceeded: the *armoire* which leaks fluid, the church under whose door is buried the corpse of Don Juan, the confessional which the priest is dragged out of, but most of all, the body. Objects are not fetishes, displacements or condensations of unconscious desires, but momentary crystallizations of a movement of contagion. Bataille's conception of space, of movement and of objects is informed by a cosmology of circulation, while Breton's is informed by the pseudo-science of magnetism and is attracted to the occult, recognizing the importance of mediums, for example.

What currents inform the Rue Blomet? What are the aesthetic terms of reference of the Rue Blomet circle? As an aesthetics of relation, movement, contagion, its general thrust can be described as an *exploration of the line*, as opposed to the fixed point. And this exploration is a development not of Surrealism but of Cubism. Cubism forms the immediate milieu out of which the movements of the 1920s emerged, while Surrealism is a parallel movement, to which Masson, Leiris *et al.* were introduced only in 1925. Their output is perhaps more pertinently seen, then, not in terms of an offshoot or a dissidence from Surrealism, but as a development from Cubism. This emergence is at once visible in terms of the network of groups, acquaintances, friendships and contacts that characterize intellectual and artistic Paris at this time, and in conceptual terms, the Surrealists, and the Rue Blomet, and the Rue du Chateau groups posing themselves often in opposition to Cubism, but always with reference to its fundamental principles. The visual art of both Masson and Miró develops out of or in opposition to late Cubism, the work of Juan Gris in particular as far as Masson is concerned, rather than in response to the more specifically 'Surrealist' form of visual art of de Chirico, which Dalí and Magritte will develop. It is therefore desirable to look not to Surrealism for the currents that inform the Rue Blomet or which it develops in opposition to, but to the network of writers and artists which forms the (late) Cubist epoch. Historical assessments of the period identify three 'Cubist poets': Apollinaire, Jacob, Reverdy.

After the death of Apollinaire, Reverdy's review *Nord-Sud* provided a kind of platform for the emerging generation, and Reverdy's writings on poetry and on art were to prove influential for Breton in his elaboration of the theory of the Surrealist image.[9] Reverdy's poetry will develop from Mallarmé in spacing out the poem on the page and proposing the word or the phrase, outlined against the white of the page, as a concrete 'thing' as well as an idea. And this spatialization also implies a network of lines forming a series of relationships that replace the fixed forms of verse or that of syntax. Reverdy will write: 'Dans ma tête il n'y a que des lignes, un faisceau de lignes.' It is in this exploration of the line that Reverdy can be called a 'Cubist' poet, in addition to his importance as a writer on Cubist painters and his friendship with Braque, for example. But Reverdy's response to the network seems to lead more to Surrealism than to the aesthetic of the Rue Blomet, for the proposal of the word as 'thing' is accompanied by a privileging of the imagery of stone and of metal. The word, become thing, is not a permutative phoneme in an unstable network—a resistance to this instability is maintained through the mineralization of the word. The network is resisted through a fixation, a magnetization which Surrealism will inherit.

Max Jacob will be closer to the Rue Blomet group in a number of ways. Jacob's Protean and farcical attitude to 'serious' literary discourse would have made any pretension to a status as theorist or as founder of an aesthetic impossible. Yet Jacob's role is crucial as an instigator of 'rencontres', a catalyst of friendships and interactions, and as a mentor. It is

---

[9] Breton cites Reverdy in the famous passage of the *Manifeste du Surréalisme* (Paris: Gallimard, Coll. 'Idées'), 31: 'L'image est une création pure de l'esprit. Elle ne peut naître d'une comparaison mais du rapprochement de deux réalités plus ou moins éloignées . . .'

chez Max Jacob that Leiris, for example, would meet André Masson, that Masson would meet Artaud, either at the Benedictine monastery of St Benoist where Jacob had taken refuge or at his 'salon' at the *Savoyarde* café in Montmartre. A network of acquaintances and friendships is generated around Jacob, which features the novelist Marcel Jouhaundeau, the dramatist Armand Salacrou (early close friends of Leiris), Roland Tual, Georges Limbour, Leiris, Masson and Miró, Artaud, Jean Dubuffet. Jacob's prominent position as an ally of Cubist painters such as Picasso, Braque, Derain or Juan Gris links the Rue Blomet circle, whose cohesion Jacob almost single-handedly produces, to the Cubist legacy. Jacob, through his ludic and irreverent attitude was also a tutelary figure for the future Surrealists—Breton, Aragon and Soupault—until the arrival of Tzara in Paris in 1919 catalysed the spirit of revolt and innovation towards the more 'serious' irreverence of Dada.

Apollinaire's influence on the generation of the 1920s is at once more vast and more imprecise. However, three aspects of this influence can usefully be identified: first, as a poet, with *Alcools* (1913) and *Calligrammes* (1918) Apollinaire explores the dynamic and visual qualities of language in a manner which I would also like to qualify as an *exploration of the line*. His verse may be seen as a continuation of the exploration inaugurated by Mallarmé's *Coup de dés*—an explosion and fragmentation of verse form, visually and conceptually, leading to a more mobile and fluid dynamic in poetry. I briefly address Apollinaire's role in the history of this literary exploration further on in this section. Its principal features are, firstly, an open-endedness of the line of verse, which, in *Calligrammes*, becomes so fluid as to seek to escape 'off the page', as it were, but is held there by the figure of the object it seeks to represent. This is linked to the second aspect of Apollinaire's influence— his role as a writer on art and as a spokesman for various movements in the world of art. His *Médidations ésthetiques: Les peintres cubistes* is at the same time an impassioned plea for a *new art* ('La vérité sera toujours nouvelle'[10]) and a descriptive assessment of mostly Cubist painting. But the ambiguity and the untheorized nature of Apollinaire's writings on art make him at the same time a champion of Cubism, Futurism, Orphism *and* a godfather for Surrealism, such that it is his role as a champion of the *new* as such which is perhaps most significant. It is due to his emphasis on the specific importance of the modern that Apollinaire is influential, not only on the Cubist movement, but also for Surrealism and the groups in its margins, for painters like Masson and Miró whose work emerged out of Cubism.

This appeal for modernity across the arts was not without a certain radicality due to the rhetorical force of its utterance. In this Apollinaire is not unlike Jacob, or perhaps, more fundamentally, Alfred Jarry, in whom the call for the new is delivered with a certain destructive humour. The personality, if 'personality' is taken to mean the *rhetorical force of utterance*, if not the thought, of Apollinaire or of Jacob is their most telling legacy for the next generation. Breton's *Anthologie de l'humour noir* of 1939 retrospectively canonizes Apollinaire (but surprisingly not Jacob), in this light, along with Sade, Ducasse, Rimbaud, Roussel and Jarry.[11] The term 'humour noir' for which Breton constructs a lineage can thus be pertinently associated with the rhetoric of what Jean Paulhan would later call 'la terreur dans les lettres'.[12] Sade, Ducasse, Rimbaud and Jarry, Roussel,[13] Apollinaire and Jacob all propose a parodic

---

[10] Guillaume Apollinaire, *Méditations esthétiques: Les peintres cubistes* in *Œuvres en prose complètes* II (Paris: Gallimard, 1991), 8.

[11] Cf. André Breton, *Anthologie de l'humour noir* (Paris: Pauvert, 1966).

[12] Jean Paulhan, *Les fleurs de Tarbes: la terreur dans les lettres* (Paris: Gallimard, 1941).

[13] Roussel is perhaps the most immediate and proximate figure of this lineage for the context of the Rue Blomet. As a friend of the Leiris family he visited 45 Rue Blomet (in 1924) and was acquainted with Masson. Roussel's *Locus Solus* was among Bataille's library loans from the BN in 1925. The process through which Roussel constructed his textual machines was not to be revealed by him until 1933, in the posthumous *Comment j'ai écrit certains de mes livres*, but Leiris's texts and *Histoire de l'œil* already show the functioning of a similar process, whereby the narrative is generated by the 'fortuitous',

destruction of the edifice of 'belles lettres' and as such are important figures for the generation of the 1920s. Bataille's *Histoire de l'œil* and the *Documents* articles evidently share in this destructive, parodic 'humour'. If Dada, which also inherits this legacy, is, for Bataille, 'pas assez idiot'[14] it is perhaps because it is not destructive enough. Not destructive enough not to have resulted in the artistic movement that was Surrealism. As such, Ducasse, Sade, Roussel or even Duchamp are more important figures to relate to Bataille than Jacob, Jarry, or Tzara.

The rhetoric of the 'épatement de la bourgeoisie' which is evidently common to the Rue Blomet group, to Bataille and to the Surrealists brings into play the importance of the obscene and erotic. They are as much elements of a desire to shock and to transgress as they are a pathology. As a writer, editor and collector of erotica and as a major instigator of the modern reputation of Sade, Apollinaire plays a crucial if problematic role, as important for the Surrealists as it was for Bataille. The re-publication of Apollinaire's erotic text *Les onze mille verges* in 1931, with a preface by Aragon, published by René Bonnel in the same collection as *Histoire de l'œil* and *Le con d'Irène* shows that despite the differences, the tradition of erotic or pornographic writing is somewhat independent, forming its own context. In this context, Apollinaire is important also as the editor of a series, entitled *Les maîtres de l'amour*,[15] which represented erotic texts from various centuries and cultures, including a selection of the work of Sade. However, Masson recalls that Apollinaire's selections from Sade, and the collection as a whole, were so 'inoffensive' as

---

but not aleatory, slippage from one set of signifers to another. Beyond this formal parallel, the *content* of Roussel's works would also feature a thematics of torture, which I consider on p. 123–4.

[14] Cf. Masson, 'Le soc de la charrue', in *Critique*, 195–6: ' "*Dada?* pas assez idiot", c'est en ces termes que Georges Bataille conclut notre premier entretien (Michel Leiris l'avait amené dans mon atelier de la rue Blomet, je lui demandais son opinion sur ce mouvement déjà extenué', 704.

[15] See Apollinaire, *Œuvres en prose complètes* III (Paris: Gallimard, 1993).

not even to attract the attention of the vice squad (*Brigade mondain*),[16] illustrating to what extent the interest in the erotic is as much conditioned by a desire to transgress in relation to the law as by individual sexual tastes or proclivities. The main source of works by Sade and information about Sade at the Rue Blomet was Roland Tual, who frequented a nearby bookseller, Georges Heilbrun. The Rue Blomet thus read an edition of the *120 journées de Sodom* published by a Dr Eugene Dühren in 1904 (the pseudonym for Dr Ivan Bloch), which Gilbert Lely reports was full of errors, and a study by Dühren, *Le Marquis de Sade et son temps* (1901) translated from the German, which Apollinaire refers to in his collection. Maurice Heine's definitive editions of the *120 journées* and of *Justine* would not appear until the early 1930s.

In these terms, the transgressive eroticism of Rue Blomet (Masson would also provide some illustrations for an unpublished edition of Sade's *Justine*) and the destructive 'humour noir' which characterize it are distinct from the less systematic eroticism and humour of Apollinaire, Jacob or Jarry. If Bataille's work is systematic, however, it is not with reference to a theory or a system: it is systematic in the destruction of its own form, systematic in the destruction of the idea of 'the work of art', or of 'literature'. In this sense its tutelary figures are Sade, Ducasse, Roussel and Duchamp—and Paulhan's notion of 'Terreur' is pertinent as a description of a cultural movement whose aim is to produce a desublimation of art, a destruction of form. Bataille's 'le rire' is 'blacker' than the 'humour noir' of Breton, while Apollinaire's eroticism is *à la mode rose*, to Bataille's *mode noir*.

Apollinaire's *Les onze mille verges*, however, does not necessarily fit into the sanctioned and identifiable eroticism of the *Les maîtres de l'amour* collection. Its author is identified only

[16] Masson on Sade: 'il n'y avait guère que les morceaux choisis par Apollinaire, fragments si inoffensifs qu'ils ne lui attirèrent même pas les foudres de la *Brigade mondaine*, ces inénarrables archanges moustachus de la Pudeur', *Ecrits*, 78.

by the initials G.A., the withholding of the name suggesting that the text does fall into the tradition of those erotic texts which transgress the law and whose authors and publishers desire to defend themselves against its prosecution. It provides a precedent for *Histoire de l'œil* in that the movement of the narrative is subjugated to the desire to construct certain sexual *tableaux*, as in Sade's writing, and it continues the legacy of Sadean eroticism through the association of eroticism and of desire with evil, pain and death. In parallel with the more structured and systematic textual operations of Roussel, the text is generated by an association of signifiers: Mony Vibescu, the hero, swears that if he does not deflower 'onze mille vierges' he should be punished by 'onze mille verges' (whips), and at the end of the text, in the midst of the Russian-Japanese conflict, this is how he meets his end. The punning contraction—'vierges' to 'verges'—is similar to Roussel's 'process' whereby a text would be generated by the desire to generate a homophonically identical but semantically different replica of an initial sentence. However, in Apollinaire's text, this process is subordinate to the construction of erotic *tableaux*, the production of different sexual permutations, while in Roussel, in Bataille, and in Leiris, the level of the signifier, or of the level of materiality as concerns the objects of *Histoire de l'œil*, is primary. *Les onze mille verges* thus provides Bataille with a pretext which he will *transform* in *Histoire de l'œil*, a textual material which will be used and operated upon, much in the same way as Duchamp takes objects as 'ready-mades' to subject them to a transformative, desublimatory process. This operation is less a production of 'art' than it is a transformation of art which destroys it, desublimates it, brings it back to the material level, back to the body.

Cubism, taken both as a reference to the specific art forms that go under that name (the painting of Picasso, Braque *et al.*) and as a general culture (in the sense that Apollinaire can be referred to as a 'Cubist poet'), thus informs the generation of the '20s, but this 'informing' has a specific character—it is a culture which will be transformed, reversed, desublimated.

If we look at the work of the Rue Blomet artists who were in close proximity to the writers Bataille and Leiris—Masson and Miró—while always bearing in mind the difficulty and difference of the relation between visual and linguistic form, some light will be shed on the culture that partly produces *Histoire de l'œil*. And the distinction of the work of Masson and Miró, for example, from both Cubism and Surrealism will also identify a specific current in modernism which *Histoire de l'œil* is also part of.

Cubism was to an extent a development of the exploration begun by Cézanne, a movement towards the identification of an ideal, abstract, sublime form. Masson and Miró's art develops out of Cubism, but in the opposite direction—instead of working from the particular towards the abstract, their painting does not move from the abstract to the particular, but breaks down form and figure into a material flux and heterogeneous ground. In the '20s, three distinct types can be identified in the art of Masson.[17] The first are drawings and paintings which depict the human figure, distorted but never so much as to break the outline of the figure and metamorphose into something else. The human figure is not a metaphor. The content of these drawings and paintings is often erotic or violent; lesbian orgies in the style of Rodin, or depictions of massacres. In the second type, post-Cubist drawings influenced by the work of Juan Gris, there is a movement from the abstract towards the particular: images and forms emerge out of an indistinct flux. In the third type, the automatic drawings which Masson began experimenting with before meeting Breton, a similar conceptual movement, from the abstract to the particular, is operating: images emerge out of the movement of the hand in the process of drawing.

Masson's position is thus pivotal in the trajectory of this aspect of modernist art, for, although Breton and Soupault's *Les champs magnétiques* had been published in *Littérature* in

---

[17] Cf. David Sylvester, 'Anatomy of this exhibition', in *André Masson: Line unleashed* (London: The South Bank Centre, 1987).

1919, Masson 'invents' the technique of the automatic drawing which will be theorized by Breton in the 1924 *Manifeste du Surréalisme*, and become one of the conceptual cornerstones of Surrealist theory. The movement from the abstract to the particular, which reverses the Cézannian principle of deriving abstract forms from the representation of particular objects, will also give rise to the 'biomorphic' Surrealism[18] which is crucial for the development of the type of abstract expressionism practised by Arshile Gorky and Pollock. It is the representation of the erotic or the violent as a movement against the figure, breaking out of its contours, which may provide the juncture of these two strands, for what is revealed in the automatic drawings is, as Masson puts it, an 'interior tumult', the expression of unconscious drives and pulsions which through the introduction of Freud into France was attaining a theoretical currency at the time. And 'biomorphic' Surrealism (although the use of the term Surrealism is perhaps misleading, for no 'higher' reality is proposed) is prem-ised on the notion of a non-figurative representation of the corporeal, the erotic. Bataille's role here may be that of a catalyst, providing a literary representation and a lyrical theo-rization (in *L'anus solaire*) of the erotic as a parodic movement, transgressing the fixity and outline of the image and of figure. If the conceptual movement of the post-Cubist art which Masson, among others, was practising is away from the abstract and towards the particular, it is because it proposes the world of images and fixed forms as transient stases of a process of metamorphosis and fluidity which is fundamental, Bataille's *continuity. Histoire de l'œil*'s innovation is to struc-ture a literary text on a fluid line of displacement of objects; it is an exploration of the line, of the line of flight or of con-tagion which links one object to another in a movement of parody.

In a text of 1968, Leiris will write on Masson's art under

[18] Ibid., 9.

the title 'Line unleashed'.[19] I have cited Leiris's statements *en bloc* here as they translate exactly the *exploration of the line* which I deem is at stake in Masson's art: 'These drawings are not *constructed*',[20] 'It is, in effect, biology and not geometry that reigns',[21] 'Masson draws as a glider pilot must, I imagine, steer his craft: finding favourable currents ...',[22] 'Something closer to the fortuitous and fleeting side of life, the unexpected aspects of nature, than to an architectural sense or a tendency towards ceremonial'[23] (citing Masson himself), 'to make fast the absence of fixity',[24] 'André Masson's errant line'[25] (citing Gertrude Stein), 'Adventure of the line (*the* line because one would like to think there is only one—always the same line, now hiding, now reappearing), of the line, which poses a being and metamorphoses it ...',[26] 'A line that moves, that is open, not one that closes or petrifies',[27] 'Arrogant lines which only rarely deign to copy reality, lines which aim as little at stylization in accordance with an aesthetic code as at reproducing a model from the outside or the inside',[28] 'Line, more than stroke',[29] 'Line in motion, more than line endeavouring to outline a structure',[30] 'lines which are intrinsically ideas, issued forth all at once from his "virile reason in a thunderbolt" as we read in [Mallarmé's] *Un coup de dés*',[31] 'the line is more than Ariadne's thread; not only a guide through the labyrinth but the force that never for an instant stops pulling him, the land surveyor, forward in its wake',[32] 'Instead of delimiting,

[19] Michel Leiris, 'Line unleashed', in *André Masson: Line unleashed*, 52.
[20] Ibid.
[21] Ibid.
[22] Ibid.
[23] Ibid., 53.
[24] Ibid.
[25] Ibid.
[26] Ibid.
[27] Ibid., 54.
[28] Ibid., 55.
[29] Ibid., 56.
[30] Ibid.
[31] Ibid.
[32] Ibid., 57.

they de-limit',[33] 'Lines that create, not lines that describe',[34] 'Seeking, questioning lines that go forward to discover where all of them will have gone together',[35] 'lines that will be anything at all—nerves, arteries, veins, antennae, waves, fibres, furrows, lodes,—rather than demarcation lines or frontiers'.[36] In Leiris's masterly assessment, Masson's line is not a limit or a frontier, a structure or an architecture, but a force which moves through and across any frontier or structure—it is a line of transgression, the same line which forms the backbone of *Histoire de l'œil*.

Post-Cubist artistic practice is 'against architecture', says Juan Gris, posing his art against that of Cézanne, which moves towards the postulation of abstract, ideal forms. The strand of modernism which Bataille's *Histoire de l'œil* is part of, which also includes the art of Masson, of Miró, and, later, of Bellmer, has its conceptual roots in an earlier critique of Cubism, which nevertheless retains its infrastructure. Dada, which specifically opposed itself to Cubism, and out of which the Surrealist movement emerged, was labelled by Bataille as 'pas assez idiot'. Not idiotic enough, since it gave rise to the Surrealist resistance to the parodic, metamorphic movement of the line. It is important to distinguish, among 'Surrealist' artists, those who sought to represent, in figurative terms, the imagery of dreams and of the unconscious, such as Dalí, Tanguy, or Magritte, influenced by the earlier work of de Chirico, from those whose art is perhaps mistakenly called 'Surrealist' in that the emphasis is not so much upon the realities which it arrives at through the processes of automatism or metamorphosis, as on those processes themselves, the processes of fluid movement and parodic contagion. Bataille's work is more apt as a theorization of this strand of 'Surrealist' art than the writings of Breton.

[33] Ibid.
[34] Ibid.
[35] Ibid.
[36] Ibid., 59.

What happens in Masson's art in the '20s, then, is a movement from Analytic Cubism to a post-Cubist practice where the direction is reversed. Emphasis is placed upon a movement away from form, against architecture. The scaffolding, the infrastructure is Cubist, but it is visually transgressed by a force of movement against structure, against the scaffold, destroying it. In Picasso's words, '[Masson] set us [the Cubists] up only to tear us apart'.[37] This process, against structure, will be accentuated in Masson's art, until it becomes almost abstract, but without structure, and resonant with the abstract expressionism of Pollock. But the more interesting moment, relevant to *Histoire de l'œil*, is the moment where the architecture is about to crumble, where the outline of the figure is ruptured.

While Masson's art therefore parallels the aesthetic of the *informe* which *Histoire de l'œil* practises, it is also specifically attached to literary practice through the illustrations for books which Masson undertakes, which include Bataille's *Histoire de l'œil*, *L'anus solaire*, the later text *Sacrifices*, Aragon's *Le con d'Irène* (also in 1928), Limbour's *Soleils bas*, Artaud's *Les pèse-nerfs* and an abortive edition of Sade's *Justine*,[38] also in 1928, and later drawings for Lautréamont's *Maldoror* and Mallarmé's *Un coup de dés*. The practice of 'illustration' characteristic of this period, for example in Breton's *Nadja*, is such that the visual representations do not have a solely 'illustrative' function. It is more pertinent to speak in terms of a conjunction of two different practices which either mutually work off each other, or develop in response to a third object. And rather than proposing that this 'third object' is an unconscious seam of imagery which informs the imaginations of both writer and artist, it is rewarding to think of the inter-

[37] Cited in William Rubin and Carolyn Lanchner, *André Masson* (New York: Museum of Modern Art, 1976), 19.
[38] Cf. Masson, *Ecrits*: 'Deux ans plus tôt [que la publication des *Infortunés de la Vertu* par Maurice Heine], l'année même de la publication des livres clandestins illustrés: *Le con d'Irène* et *Histoire de l'œil*, A.M. avait entrenu une série d'études pour *Justine*', 95.

change as a process of reading. The reading of the text by the artist, which produces the images, necessarily goes by way of a reading of the textual corpus which the text in question activates and transforms. The precise relation between Bataille and Masson, then, takes place as a process of reading of a textual corpus in which the most prominent name is that of Sade. The Rue Blomet thus provides a forum, in a sense, for the reading of Sade, and it is likely that Bataille's interest in Sade develops out of this forum, his first reference to Sade appearing in the article on pre-Columbian art of 1928. We can therefore propose that Masson's 'illustrations' for *Justine* form part of the textual corpus the reading of which produces *Histoire de l'œil*. One such illustration depicts a scene from *Justine*, showing a woman, hanging, her sex thrust out towards the viewer, in the throes of orgasm. Beside this figure, but only partly visible, sketched in, is a male figure, with one hand masturbating, and the other arm outstretched in a gesture either of pointing to the heavens, or of command. Masson writes to Kahnweiler, in 1928, that he had 'une certaine prédilection'[39] for this scene. This 'certain predilection', produced from a reading of Sade, will also inform *Histoire de l'œil*. The scene of Marcelle's suicide by hanging and her nightmare of the Cardinal who is an executioner is resonant with it, but in a more general sense the association of sacrifice, death and *jouissance* is a powerful seam of *Histoire de l'œil*. If Masson's drawings accompanying *Histoire de l'œil* are certainly more 'illustrative', in that they are produced from a reading of that text, they also draw out its mechanism by depicting the key objects (eye, egg, sun, vagina) as fixed points in a series of fluid lines. But it is also possible to say that the text of *Histoire de l'œil* itself is produced through a reading of Sade which operates via the drawings that Masson undertook of *Justine*. Thus the Rue Blomet functions as a kind of experimental laboratory, or a reading forum, which produces the text. In a sense that is at once more general and more specific,

[39] Ibid.

there are also certain points of intersection between the writing of Bataille and the art of Masson in certain key figures: the labyrinth, the minotaur, the massacre, culminating in the figure of the *acéphale* which Masson produced for the periodical of the same name. These figures are also produced as readings of a textual corpus, whether mythical or literary.

Masson is certainly the most important of the artists who frequented Rue Blomet in terms of his influence on Bataille. But it is important to resist the tendency to talk in terms of the intersubjective *influence* of one person on another, even if the force of friendship (Blanchot's *l'amitié*) is decisive. Rather than identify Masson, or Leiris, as individually important for Bataille, it is the Rue Blomet which functions as a space in which various activities and discourses took place, which produces the text. To this extent the art of Joan Miró is also a factor to be taken into account.

This accounting has been admirably proposed by Rosalind Krauss in her essay 'Michel, Bataille et moi'[40] which takes as its starting point a painting by Miró with those words written across it. The inscription of Bataille's name in Miró's painting metaphorically signals Krauss's attempt to reinscribe the name Bataille in the discourse on Miró, and more generally of modernist, post-Cubist art. The aim of Krauss's article, it seems, is to dissociate Miró from one strand of modernist art and re-associate him with another, in the context of a generalized effort to plug modernism back into the corporeal register which Bataille's writing affirms. Krauss criticizes what she sees as the 'prudishness of Miró commentary',[41] associating it with an 'occlusion of Bataille'.[42] Bataille is, she asserts, indelibly written into Miró's art, and it is necessary to engage with this. She writes against what she sees as a tendency to associate Miró with colourfield painting and a transcendent, metaphoric aesthetic which is represented by Jacques Dupin's

---

[40] Rosalind E. Krauss, 'Michel, Bataille et moi', in *October*, 68 (Spring 1994), 6.
[41] Ibid., 10.
[42] Ibid., 6.

monograph on his work, and in favour of a reading of Miró's paintings which emphasizes their fetishistic, sexual elements. She notes that throbbing genitalia enter Miró's work in summer 1924 (a few months before Bataille's arrival at Rue Blomet), and that the paintings function as 'metaphoric strings of relationships based on these organs',[43] of which the 'explosive climax' is determined by an expression of the sexual. An implicit, unstated step is the association of Miró's practice with that of *Histoire de l'œil*, objects appearing as fixed points in a series of lines, the lines which appear in Miró's art functioning as visual representations of the process of metonymic slippage operative in the text. However, if Krauss's aim is to dissociate modernist art from a transcendent horizon, and is therefore resonant with Bataille's critique of Surrealism as idealist, and if she accurately identifies (in a different text) the permutative play of objects in *Histoire de l'œil* as a desublimatory, *informe* practice, the identification of the 'expressive goal' as sexual, and of figures as fetishes, seems to resituate an origin, to reterritorialize the permutative play of objects or figures in relation to a teleology. We might recall here Barthes's insistence on the lack of explicitly sexual orientation of the metaphors in *Histoire de l'œil*, which Krauss also cites.[44] The 'explosive climax' in *Histoire de l'œil* is a series of climaxes, determined not by desire, a desire for *jouissance*, but by the intersection of objects and figures, and words. Using *Histoire de l'œil* to read Miró's work would thus emphasize the permutative play of figures, the *exploration of the line*, rather than the fetishistic aspects of the figures. Miró would then appear as an artist of the Rue Blomet whose work evolves out of the same crucible as that of Masson, in response to Cubism, as an exploration of the line. We could thus propose Miró's art as something the reading of which has an effect in the production of *Histoire de l'œil*. The visual representation of different figures, linked by lines, permuting the attributes

[43] Ibid.
[44] Ibid., 10.

of those figures and articulating them with the register of the sexual and corporeal is a process which *Histoire de l'œil* will develop and parallel.

Answers to the question *where*: Bataille's initial intervention is thus *from within* the institution, and evidently conditioned by a number of biographical events which we will consider later, in the context of a discussion of the 'autobiographical' section of *Histoire de l'œil*. It is *into* a society of groups in the margins of the Surrealist movement in 1920s Paris. It is *by way of* an original vision of movement and of thought which is deemed 'pathological', and is *informed by* an aesthetic whose character I have outlined.

12. / auch

Histoire de l'œil
par Lord Auch

Le Petit, by 'Louis Trente' (Georges Bataille), published in 1943, includes a 'Préface à *Histoire de l'œil*' which reveals that the pseudonym 'auch' derived from a blasphemous contraction of the phrase 'Dieu aux chiottes' (God in the shithouse):

> Il est dans l'*Histoire de l'œil* une autre réminscence de *W.C.*, qui, dès la page de titre, inscrit ce qui suit sous le signe du pire. Le nom de Lord Auch se rapporte à l'habitude d'un de mes amis: irrité, il ne disait plus 'aux chiottes', abrégait, disait 'aux ch'. Lord en anglais veut dire Dieu (dans les textes saints): Lord Auch est Dieu se soulageant.[1]

Leaving aside, for a moment (see p. 82), the perverse nature of this preface, in a different text and fifteen years after the event, and of this revelation, by a differently named author

---

[1] *O.C.* III, 59.

who refers to himself as the same, the *name of the author* should detain us here, especially since Louis Trente signals its capital importance for the text; it places 'all that follows under the worst of signs'. 'Lord Auch', after *Histoire de l'œil*, functions as a second title. By the logic of capitalization (which Bataille would later study) through which the head, the title, *stands for* what it capitalizes, the text of *Histoire de l'œil* is placed under the sign of contraction, and, moreover, under a contraction which dislocates the paradigm of the capital, the highest possible capital of capitals: *Dieu*. This lapsus is effective both as a profanation and as a decapitation, a decapitalization which parodically undermines the capacity of the second title to function as a title, which places the text under the heading of decapitation and decapitalization, under the sign of the *acéphale*.[2] *Dieu*, moreover, is rendered as 'Lord', which, presented as a proper name, 'Lord Auch', translates it from the position of ultimate transcendence into the contingency of the English nobility and the textual network of a tradition of pornography, in which the male protagonists are titled. *Histoire de l'œil* has its own representative of that textual tradition: Sir Edmund, 'un richissime anglais'.[3] *Dieu* is no more than an English 'Lord', despite the reference to the Scriptures, no more than a textual figure from the worst of literary traditions.

The text is placed under the sign of a linguistic short-cut, a pun or a joke. The joke, Freud's *Witz*,[4] short-circuits the

---

[2] See Bataille's articles for the journal *Acéphale* which he founded with André Masson and Pierre Klossowski. Cf. also Jean-Joseph Goux, 'Numismatiques', *Tel Quel*, 35–6 (Autumn 1968 — Winter 1969). Goux studies the logic of capitalization in the context of Saussurean linguistics, Lacanian discourse on the phallus, and Marx's analysis of money, proposing 'une pensée radicalement acéphalique'.

[3] *O.C.* I, 48.

[4] The following comments from an article by Jean-Luc Nancy, titled 'Menstruum Universale', in *The Birth to Presence* (Stanford, Calif.: Stanford University Press, 1993), can serve here but to point to a potentially fruitful line of enquiry, which would investigate the echoes and resonances between the theory of the *Witz*, not only Freud's (in *Jokes and their Relation to the*

repressive and discontinuous rationality of discourse, the division of the world into distinct objects and categories, and enables a momentary and paroxystic continuity to be glimpsed. This instantaneous continuity, or sudden condensation, to use the language of the dreamwork, is *obscene* because it disturbs the structural stability of representation, where one thing is distinct from another. It is also a shock which remains outside representation, a sudden paroxysm whose bodily expression is the laugh. As a blasphemy, a profanation, it is a formal transgression of the discontinuity of human conceptualization, which separates the 'high' from the 'low', and it may be the most consequent of such transgressions, since it forces together the 'highest' and the 'lowest' of human experiences, in a manner which *Histoire de l'œil* will exploit (see p. 105). *Auch*: the name of the author is a pun, which, in an instant, illuminates a deep continuity, a *communication*. The story is written 'in the name' of this kind of contraction, and our reading has to operate according to this programming.

For example, 'il y avait dans le coin d'un couloir une assiette de lait destiné *au cha*t'.[5] We are invited to read the text as a tissue of puns and linguistic jokes which provoke anxiety and obscenity. We find here the contraction in the name of which the text is written (the name of the author), coincidentally,

---

Unconscious) but also that of the German Romantics, and Bataille's punning, indeed the whole tradition of punning and wordplay in French literature since Surrealism: '*Witz* is barely, or only tangentially, a part of literature; it is neither genre nor style, nor even a figure of rhetoric' (248); '*Witz* does not *hold* the positions that theory—any theory whatsoever—might want it to occupy' (249); 'In the state in which it came the closest to being a true concept—with the Romantics—*Witz* most generally designated the union, the mélange (or the dissolution) of heterogenous elements' (251). Particularly in relation to this last comment and its note which suggests the indebtedness of much contemporary theoretrical discourse to the *Witz* and moreover to the Romantic theory of the *Witz*, we might identify Bataille's notion of *communication*—the transgression of discontinuity—as a fundamental influence on these discourses, and raise the question: to what extent is Bataille a Romantic, indebted to the tradition of German Romanticism?

[5] *O.C.* I, 13.

in the lead-up to the text's first pun, Simone's 'Les assiettes, c'est fait pour s'asseoir, n'est-ce pas?';[6] as readers we are *led on* into a reading of the text according to puns, linguistic slippage and obscenity. The function of the *story* as a narrative or as a (fictional) representation of a series of events in the world is already being undermined.

13. / framing

Histoire de l'œil
par Lord Auch

The simplicity of representation, of the nature of *story*, is made problematic by the multiplication of the frame. The frame, in the simplest of worlds, consists only of a few directives: the name of the author, the title of the book, the location of the events. In the case of *Histoire de l'œil* the situation of the narrative, of the story, is complicated by a number of factors: by it being already designated as a *story*, first of all—by the fictionalization of the author's name and its inscription in the logic of the text (that of punning)—by the partial location of the text in relation to a geographical and historical reality— by the second part of the text, 'Coïncidences', in which the author describes the psychological and psychoanalytical deter- mination of the text, after the event. One effect of this is to provoke the question: what kind of text is this? what is its genre? The question the text provokes is the question of genre. The title already explicitly tells us that it is a story, an '*his- toire*', but adds another, very slightly displaced, generic determination further into the text, signaling its 'first part' ('Première partie') to be a, or the, 'récit' (narrative).

[6] Ibid.

Histoire de l'œil
par Lord Auch
Première partie
Récit

We are always justified to a certain extent in reading the linear
order of titles as a hierarchy. The immediate implication of
such a reading is that the story, *Histoire*, includes both the
'récit' and the second part, 'Coïncidences' ('Réminiscences' in
the later version). The story would be a larger entity than the
narrative (*récit*), including also a consideration of how the narrative was generated from events in the life of the writer. Of
course, if we obey the hierarchy imposed by the order of titles,
or headings, the second is 'Lord Auch', situating the author as
an entity within the fiction (the story), the identity of the real
author remaining unknown, or at least ambivalent. But this
hierarchy is evidently going to be disturbed by what occurs
within the text; the order of words does not obey a pure linearity. The word 'par' already situates the author half in and
half out of the fiction, and the second part of the 'histoire' will
go on to complicate the structure by first of all introducing a
different 'je', the author as opposed to the unnamed 'je' who
narrates the 'récit', and then throwing doubt upon this difference: 'Je croyais ainsi, au début, que le personnage qui parle
à la première personne n'avait aucun rapport avec moi.
Mais . . .'[1] 'Coïncidences' then establishes a certain association
(explored later, in 24. *Cure*) between events in the life of the
author, and the *writing* of the events of the narrative. In other
words, it draws attention to the process of the writing of the
text, its unconscious determination, after the event, and in
doing so doubles the fictional world of the *récit* with a scene
in the real (the scene of writing) which frames it. The framing
of the fictional *récit* by the real is, moreover, exploited by the
situating of the text in relation to a real event, in historical
time and geographical space: the bullfight in Madrid, on 7 May

[1] Ibid., 73.

1922: the event of Granero's death. The text of 'Coïncidences' insists on this fact; Granero is a 'personnage réel', the corrida is an event 'à laquelle j'ai réellement assisté'[2] (it is the writer of 'Coïncidences' speaking). And it is significant that it is in Spain that the 'reality' of the events in the narrative is located, since 'Spain' functions, in Bataille's writings, as the scene of the 'real', of an 'acting out' in relation to the 'fictional' and imaginary space of France.[3] The reality of the event around which the fictional *récit* weaves itself, and its chains of objects, further complicates the simplicity of the representational structure (of *mimesis*) and frames the fiction, as if from within. Mimesis is complicated as a simple structure, as a structure of the re-presentation of a scene (real or imagined) and proposed as a series of repeated dislocations, of dislocated frames. If, as Bataille suggests in the preface to *Le bleu du ciel*, 'tout homme est suspendu aux *récits*',[4] the suspension of the *récit* (its fictional status, as a result of the suspension of disbelief) is undermined and problematized by the multiplication of the frame.

So: what kind of text is this? What genre of text is this? The question assumes one can say that a genre *is*. We have been unable to say what it *is*, and this because the temporal structure of generic determination always insists not on the evidence, in the presence of the text, of the signs of genre, but on its *coming after* a certain law or code of genre. We can say what the text *transforms*. If *genre* is always a concept which the text transcends or transgresses, by always coming after the designation of a kind, of a *genre*, one can identify what *genre* the text comes after. Or what *genres*: it is always possible that a text comes after a number of genres, obeys and disobeys the

[2] Ibid., 74.
[3] In *Le bleu du ciel* the narrator's political instability is *realized* in Barcelona. One might posit this relation to Spain, moreover, as a trope which is repeated in French writing of the twentieth century, in Montherlant and Malraux, for example—Spain functioning as a scene of political or physical action, displaced southwards from France, which refuses it.
[4] *O.C.* III, 381.

laws of a number of genres. *Histoire de l'œil* alerts us from the start of its desire to obey and disobey the laws of the genre: *histoire*, but, in that the *récit* and the second part, which for the moment remains without a named genre as such, are both contained within and under the title *histoire*, *histoire* appears as something like the *genre of genres*: *histoire* would designate the simple fact of being related to, being a matter of, being about. *Histoire de l'œil*: it's all about the eye, it's a matter of the eye, this is a story that touches upon the eye.

By saying 'it's about the eye', the title points to the act of reading and therefore the *transformation* of other genres, other texts. Identifying what genres *Histoire de l'œil* transforms is to point to what happens in the process of its writing, and to suggest that it is a more complex operation than the 'simple' representation of sexual fantasy. The confessional, the Gothic novel, pornography, the adventure story: elements of these genres will be picked up in the reading which follows, figures which *Histoire de l'œil* takes from a previous body and transforms in its own crucible.

14. / narration

In narration, time is organized into a series of events which can be distinguished from each other. This is how it can be made sense of. Time is chronology, organized into a line connecting a series of points, a series of moments that were future moments, are now present moments and will be past moments. Narrative requires the representation of affects, perceptions or sensations as events. What does not allow itself to be represented as an event, and thus organized as a moment in a narrative sequence, is a shock or affect which remains unrepresentable, will not allow itself to be represented, because it is immediate. This tautology suggests the difficult and unmediatizable nature of the immediate shock, the initial violence or instantaneous paroxysm, which Bataille's *continuity* produces in discontinuous reality. That is to say that the narrative of *Histoire de l'œil* hides, covers an unrepresentable, unnarrativizable *thing*, an *obscene* vision which forces its way through and ruptures the narrative chain at specific moments. The narrative of *Histoire de l'œil* is one which is *interrupted* at a number of points by an instant (an instant which nevertheless remains outside the capacity of narrative to organize into a chronological chain) which has something of the quality of the sublime. *Histoire de l'œil* stages the interruption of the *récit* by the obscene.

The interruption is, however, determined by a different way of organizing reality. Instead of thinking of a series of events linking discontinuous moments, time is in some sense condensed, squeezed together towards an explosive instant through the transgression of the frontiers and limits which mark this discontinuity. The being present of the present moments is contaminated; it leaks out into a different moment (Proust's experiences of involuntary memory function in the same way). This leakage operates by way of an associative chain of objects and words, which are superimposed on one another. The fixity of the attention on one object at one particular moment leaks into other objects and thus into different moments. The organization of time into a horizontal stream of successive present moments is contaminated by a 'vertical' principle of continuity between objects and between moments. The narrative is thus threatened by a centrifugal pull towards a single instant, a singular object which would be the annihilation of all objects in the hole of their own total contiguity. The ultimate principle of *Histoire de l'œil*, the force which threatens its readability but also determines it, is a maelstrom, a hole.

The function of the *récit* of *Histoire de l'œil*, therefore, is not primary, but subordinate to a non-narrative, atemporal chain of association between objects, whose character I consider further on. Nevertheless, the permutative association of objects is applied to and set against a narrative, which, while it is interrupted and undermined by the hidden logic of the associative chain, is also determined by other considerations. The *récit* has a spatial and temporal organization which nuances the pure action of permutation.

The *récit* of *Histoire de l'œil* is shared between two distinct geographical 'spaces', which divide the text into two parts, the first (Chapters 1–8) taking place around the village of 'X', presumably in southern France, and a nearby asylum. The second part (Chapters 9–14) leads on chronologically from the first, guiding the characters on an identifiable itinerary down through Spain (Madrid–Seville) to end in Gibraltar with the adventurers' departure on a yacht, and woven around two distinct locations, the arena in Madrid and the 'church of Don Juan' near Seville. While the geographic location of the text imposes a kind of closure, programmes an end (the narrative ends when the trio reach Gibraltar and leave Spain, on a yacht hired by Sir Edmund) this end is not definitive, since the adventurous trio simply come to the end of the land, of Europe, and continue their voyage of discovery on sea. And while there is a semblance of narrative logic—the transition between the two parts of the narrative, its two spaces (France and Spain) provided by the necessity for the characters to flee the scene of a crime—the narrative transitions from one scene to the next are minimal and parodic. The narrator and Simone escape to Spain 'pour éviter l'ennui d'une enquête policière'[1] stealing a car and subsequently enlisting the aid of 'un richissime Anglais, qui lui avait déjà proposé (à Simone) de l'entretenir'.[2] The text spends very little time on narrative transition from one space to the other, does not concern itself with establishing the chronology as *vraisemblable*, or with descriptions of its locations other than those taken up in the logic of the text's structure of images. Elements which fall outside determination by the chains of association are strictly dispensable, token

[1] *O.C.* I, 48.
[2] Ibid.

gestures to narrative continuity. The narrative *vraisemblable* is parodied, as is evident in the final episodes of the text, where the narrator, Simone and Sir Edmund are again fleeing the scene of the crime (the murder of the priest):

> Deux heures après, Sir Edmund et moi décorés de fausses barbes noires, Simone coiffée d'un grand et ridicule chapeau noir à fleurs jaunes, vêtues d'une grande robe de drap ainsi qu'une noble jeune fille de province, nous quittâmes Séville dans une voiture de louage. Nous changions nos personnages à l'entrée d'une nouvelle ville. De grosses valises nous permettaient de changer de personnalité à chaque étape afin de déjouer les recherches policières. Sir Edmond déployait dans ces circonstances une ingéniosité pleine d'humour: c'est ainsi que nous parcourûmes la grande rue de la petite ville de Ronda lui et moi vêtus en curés espagnols, portant le petit chapeau de feutre velu et la cape drapée, fumant avec virilité de gros cigares; quant à Simone qui marchait entre nous deux et avait revêtu le costume des séminairistes sévillans, elle avait l'air aussi angélique que jamais.[3]

Narrative continuity is presented as 'risible', reduced to a caricature with a grotesque, excessively theatrical nature. Nevertheless, the *récit* of *Histoire de l'œil* borrows the minimal structure of the adventure or crime story: a crime is committed by a couple, they flee, are joined by an accomplice, their violence becomes more and more pronounced, they murder an 'innocent' victim, and flee again. The minimal structure of this genre is borrowed, but without any attempt to elaborate it, to provide it with motive or moral—the authorities are more or less absent from this parodied Bonnie and Clyde story. The genre of the adventure story is used solely as a scaffolding for the associative chain of objects and the staging of the obscene in a number of 'scenes'.

This kind of transformation of the genre of the adventure story—only minimally and parodically present—is not, however, a strategy limited to *Histoire de l'œil*. It is a transformation that is already operating in another genre: the erotic

---

[3] Ibid., 69.

or pornographic novel. Apollinaire's *Les onze mille verges* plots its hero Prince Mony Vibescu's itinerary across Europe to Vladivostock, constantly in flight from hideous crimes.[4] Narrative is reduced to *flight*, the movement of the hero—but it is less conditioned by the psychological motif of fear or guilt than by the necessity of moving from one scene to the next. Without narrative, without *flight*, the text would consist solely in the representation of various scenes and the successive alteration of sexual permutations, much as the novels of Sade privilege the 'montage' of lubricious scenes. The narrative of flight also enables the staging of the sexual to adopt elements of its location in its representation of the erotic. In Apollinaire's text, therefore, the erotic becomes more and more exotic, a strategy common to contemporary erotica, Darsat's *Emmanuelle*, for example. In *Histoire de l'œil*, in a more complex mode, it becomes more 'Spanish'—closer to the arena of bullfighting, to the church of Don Juan—in other words to specific locations which are determined by concerns other than those of narrative or sexual permutation—the associative chain of objects and the mythic or intertextual strategy of the text. One might, however, propose that the movement of such narratives from France to the extremities of Europe, which occurs in both Apollinaire and Bataille's texts, is conditioned by an exceeding of the possibilities of sexual permutation offered in Europe, a movement towards the exotic. In Bataille's text, however, this movement is interrupted and this generic strategy reduced again to a parodic minumum.

Narrative, therefore, is *risible*. It is merely a scaffolding for the staging of the obscene, a minimal structure there to be *interrupted* by the paroxysmic moment of orgasm, and undermined by the associative play of objects and words. Narrative as the representation of successive events, determined, according to structural analysis, by the search for truth or the unveiling of a secret, is replaced by different kinds of organization, different kinds of narrative.

[4] Cf. Apollinaire, *Œuvres en prose complètes* III (Paris: Gallimard, 1993).

While the two distinct spaces of the *récit* seem to divide it into two separate and consecutive parts, this relation is undercut by what we might call the 'primary narrative'—the association of objects and words. Seen in this light, a more precise relation between the two parts becomes evident. The first part stages the initial, originary experience of the object or of associations between objects, or words, an experience which causes a sexualized anxiety. The second part of the text, set in Spain, will function as the deliberate *application* of the first set of associations to a new set, their *intersection* with other objects and other chains of association, reactivating and extending the sexual anxieties of the first section. The fact that the first section is set in an unidentified oneiric topography and the second in a geographically 'real' space is also significant. Spain, as I have proposed, functions as the 'scene' on which transgression is acted out. To take an example, to the proximity established punningly by Simone in the first part, of her vagina with the saucer (assiette), is added, in the bullfight scene, the elements of the bull's testicles. And, to take another example, the text as a whole is organized around the decisive moment of Granero's enucleation. The paroxysmic moments of the text are those at which the potential, imaginary associations already suggested are acted out: the gouging of Granero's eye is the literal (as opposed to metaphorical) migration of the eye out of its socket, the beginning of its downward journey. The associations or objects with *the eye* which are produced earlier in the text function solely as metaphors; the eye in its socket *cannot*, in the real, act as an object on its own. Granero's enucleation *enables* the liberation of the eye from its socket, the *realization* of the metaphor of the eye's mobility, which in turn will be *realized* upon the body of the priest in the final scene. Relations between the two parts of the *récit* can thus be expressed in terms of *application* or of *realization*.

To some extent *Histoire de l'œil* also follows a strategy which we might think of in psychoanalytic terms as the manifestation in the real of repressed anxieties. The first sentence focuses attention on the productive nature of *anxiety*: 'J'ai été élevé très seul et aussi loin que je me rappelle j'étais angoissé par tout ce qui est sexuel'.[5] The text is essentially concerned with this *anxiety*, and with its manifestations and its effects in the real. This anxiety is in response to an instantaneous glimpse of a continuity, it is the delayed effect on the psychic apparatus of the shock of the pure unmediated and indistinct existence of the outside, of a total ground unlocated by any figure. The anxiety in response to the continuous ground leads the characters to stage the association of distinct objects, to represent continuity by acting it out. The *histoire* of *Histoire de l'œil*, then, is as much a psychic parable as an experimental, structurally subversive text—it tells the story of two characters who are led to various actions through an anxiety induced by the instantaneous vision of an indistinct continuity.

The story of the eggs can serve as a useful example of this. After the erotic play with eggs between the narrator and Simone in the first part, they resolve not to speak of eggs henceforth:

> Mais il faut le dire ici que rien de semblable n'eut lieu depuis entre nous et, *à une exception près*, il ne fut plus question des œufs dans nos conversations: toutefois si par hasard nous en apercevions un ou plusieurs nous ne pouvions nous regarder sans rougir, l'un et l'autre, avec une interrogation muette et trouble des yeux.[6]

And further on:

> Nous avons toujours évité, Simone et moi, par une sorte de pudeur commune, de parler des objects les plus significatives de nos obsessions. C'est ainsi que le mot *œuf* disparut de notre vocabulaire ...[7]

Eggs are in fact banished from the second part of the *récit*, and do not appear, *as eggs*. The anxiety which the earlier

---

[5] *O.C.* I, 13.
[6] Ibid., 39 (Bataille's emphasis).
[7] Ibid., 40 (Bataille's emphasis).

'divertissement' had induced is explicitly displaced on to another object:

> On verra d'ailleurs à la fin de ce récit que cette interrogation ne devait pas rester indéfiniment sans réponse et surtout que cette réponse inattendue est nécessaire pour mesurer l'immensité du vide qui s'ouvrit devant nous à notre insu au cours de nos singuliers divertissements avec les œufs.[8]

Why should amusement with eggs induce anxiety? Why should games with eggs open a 'vide'? The solution lies, I think, in the notion of a 'vide' underlying the text as a whole, and opening up at certain moments in the text where the association of objects 'correlate' with it. Further on I will look at the proposition of an original, determinative figure at the heart of *Histoire de l'œil*, which is the source of this anxiety—this is the *figure of the eye*, or the eye as figure.

The second part of the *récit* follows the itinerary of the psychoanalytic cure or analysis, releasing the repressed obsessions and anxieties into action, with the difference that the textual 'analysis' or cure, parodied in Simone's 'confession' in the church of Don Juan, does not lead 'back' to social stability and responsibility, but instead releases the blocked-up anxieties into a murderous acting out. Discourse (analysis) does not function as the representation and therefore the sublimation of aggressive drives; anxiety is not ironed out in discourse, it cannot be spoken and remains locked into a fixed mutual gaze. Bataille's text is the representation of the return, 'in the real' of anxiety non-sublimated in discourse.

The narrative structure of the text thus corresponds to a broadly psychoanalytic pattern. But with significant differences. *Histoire de l'œil* is not a 'closed' text in the sense that psychoanalysis returns the analysand to normality and sociality, closes the analysis through the end of the transference; its power continues to have an effect even after the final, 'risible' sentence:

[8] Ibid., 39.

> Le quatrième jour, l'Anglais acheta un yacht à Gibraltar et nous prîmes le large vers de nouvelles aventures avec un équipage de nègres.[9]

The *récit* ends with a continuation. Moreover, as I have already suggested, the obsessions and chains of association do not lie 'underneath' the text, there to be discovered by textual exegesis, and seemingly belonging to an unconscious interiority. The text explicitly brings its own functioning to the fore, through a narrative voice which takes the place of the analyst, so to speak. As well as the distinct section 'Coïncidences', set apart from the *récit*, where the 'author' analyses the fiction in relation to his own past experience, within the text the narrator acts as a analyst and commentator, pre-empting any position of interpretation on the part of the reader or critic. The text enacts its own auto-analysis, establishes a metatextual, narrative voice within the narrative, and this voice is in close but also parodic and subversive proximity to that of the analyst.

As *récit*, or *histoire*, *Histoire de l'œil* is at once simple and complex. Simple if narrative is understood solely as the succession of events, as *story*. If we understand the title, *Histoire de l'œil*, however, not only as 'a story of what happened to the eye', but also as 'a story about the eye', a 'story to do with the eye', then the complexity of the text as a transformation of a psychoanalytic case history of a kind, and as a complex undermining of narrative logic by a staging of the obscene, becomes manifest. The staging of the obscene proposed here requires that we consider the text according to the logic of its associations of objects and words, as an alternative narrative which disturbs temporal organization and therefore cannot with security be named as narrative.

[9] Ibid., 69.

15. / roundness–liquidity–light

As Roland Barthes proposes, the 'story of the eye' is the story of an associative chain of objects represented by words linked to each other either through their phenomenological qualities or through their linguistic proximity, or both. The fact that the story is generated and the actions of the characters either consciously or unconsciously determined, by *formal* associations between objects or words undercuts the usual motivation of narrative by a psychology of depth or by a search for truth, an Oedipal secret. It is the formal and structural determination of the text, and of its erotic scenes, which makes of it a radical and unsettling experience of reading. It cannot be ascribed to fantasy, to the imaginary, nor simply to a will to 'épater le bourgeois', far less to a representative account of what once happened or was imagined. It can only be ascribed to form, to a certain formal strategy which *then* comes to affect, and to produce anxiety in, the characters upon its stage.

I look at linguistic associations in 16./punning. The phenomenal qualities which the text exploits in its contagious associations are those of roundness, liquidity and light. The reader can easily identify a chain of round objects in the text: the saucer Simone sits in; the plate (also an 'assiette') in which the bulls' testicles are placed later, at the bullfight; eggs ('les œufs'); the bowl of the W.C. in which Simone drowns the eggs ('la cuvette'); the bull's testicles ('les couilles du taureau'); the sun ('le soleil'); the moon ('le disque lunaire'). Close to this chain, and associated with it, are semi-round, ellipsoid or approximately round objects: the eye itself ('l'œil' and its plural, 'les yeux'), Simone's sex ('le cul'), and the stain on the sheet which Marcelle hangs out of the window of the asylum ('une tache').

It would be difficult to reconstruct any total and fixed code or structure that would act as the generative matrix or grid for the text and its images, as I proposed earlier; the text metonymically associates one element with another at such rapid speed that this tabular reconstruction is disallowed. The reader is left with an activity of structuring which effaces itself as it proceeds. What results is a sensation of vertiginousness (*le vertige*).

We might propose, however, one provisional structural line as follows: the narrator looks at Simone's sex in the saucer of milk; he subsequently looks at her pushing eggs into her vagina above the bowl of the W.C., and then at eggs imploding in the bowl; when Simone and the narrator visit the asylum in which Marcelle is imprisoned they notice a strange, 'reversed' similarity between the moon and the stain on the white sheet which Marcelle has hung out of the window. Later, in the second part, in Spain, the saucer returns, this time containing a castrated bull's testicles, which Simone inserts into her sex at the moment when the bullfighter Granero is gored by the bull and his eye bursts out ('jaillit') from its socket. In the scene in the church of Don Juan, the eye of the priest is cut from its socket and inserted into Simone's vagina. The narrator recognizes in it the eye of Marcelle.

The structure of the primary narrative of the text can thus be proposed in relation to the figure of the insertion or ejection of different round objects into Simone's sex, ending with the eye of 'Marcelle'. Such are the acts in the text which realise the potential contiguity of all of these round or semi-round objects, a contiguity which suggests a threatening transgression of the distinctness and fixity of each object. The solid outline of the object is ruptured in its displacement into the next.

This primary narrative, as I have called it, is evidently extremely simplified in this account. Liquidity, or liquefaction, provides a second chain of associations which is *applied* to that of roundness. Liquefaction is the movement of contiguity between the objects whose roundness signifies their being limited. Liquefaction transgresses and ruins those limits. Thus the text privileges not only liquids—milk ('le lait'), urine ('l'urine' or 'le pisse'), sperm ('le sperme' or 'le foutre'), blood ('le sang'), rain ('la pluie')—but also the process of liquefaction, the action of liquefying: flowing ('couler'), soaking ('tremper'), streaming ('ruisseler'), inundating ('inonder'), pissing ('pisser'), moistening ('mouiller'), spurting ('jaillir'), spitting ('cracher'), drinking ('boire') and nouns and adjectives associated with this process: the jet ('le jet'), the drop ('la goutte'), the trickle ('le filet'), the stream ('le ruissellement'), the torrent ('le torrent') the pool ('la flaque'), the stain ('la tache'), the humid ('l'humide'). At important junctures of the text, the lexicon of liquidity is metaphorically transposed on to unusual objects, for example, in the phrase which the narrator uses to refer to Simone 'kissing' his eye: 'pour ainsi dire buvant mon œil gauche entre ses lèvres',[1] or in the description of Granero's costume as 'érigée très raide et comme un jaillissement...'[2] The principal action associated with liquidity is urination. The narrator and Simone are continually urinating upon each other or upon different objects. This in itself is commensurate with the erotic nature of the text and the erotic play between the characters; its *cosmically* transgressive function results from the *application* of the action of urination to the universe as such; reality is liquefied, the real *flows*. And the symptom and signal of this liquefaction of the real is the intersection or association of the semantics of liquidity and those of light and vision, thus leading us back to the figure of the eye: the real, as what is *seen*, flows. Light is liquid. There are thus a number of superimpositions or associations of liquidity and light: the jet of urine is explicitly associated with or superimposed on to the *jet* of lightning, just as the stain on Marcelle's sheet is superimposed on the moon and then the lit window of Marcelle's room in the asylum, a hole in the darkness. At one of those moments of the text where the narrator takes on a metatextual, analytic function, he comments upon this association between liquidity and light:

> L'urine est pour moi profondément associée au salpêtre, et la foudre, je ne sais pourquoi, à un vase de nuit antique en terre

[1] Ibid., 38.
[2] Ibid., 52.

poreuse abandonné un jour de pluie d'automne, sur le toit de zinc d'une buanderie provinciale. Depuis cette première nuit à la maison de santé, ces représentations désespérantes se sont unies étroitement au plus obscur de mon cerveau avec le con comme avec le visage morne et abattu que j'avais parfois vu à Marcelle. Toutefois ce paysage chaotique et affreux de mon imagination s'inondait brusquement d'un filet de lumière et de sang, c'est que Marcelle ne pouvait pas jouir sans s'inonder, non de sang, mais d'un jet d'urine clair, et même pour moi illuminée, jet d'abord violent et entrecoupé comme un hoquet, puis librement lâché et coïncidant avec un transport de bonheur surhumain.[3]

We pass here from the register of liquids—urine and rain—to that of light, and by association with the jet of lightning, to the flash, the paroxysmic moment of orgasm, the 'hoquet', or the blink of an eye, which ruptures or punctures the discontinuous surface of the real. Furthermore, at the climactic moment where Granero's eye is ejected and Simone inserts the bull's testicles into her vagina, there occurs a radical intersection between the theme of liquidity and that of vision and light, when the sky becomes so bright, yellow, humid and heated, described as a 'luminosité éclatante mais molle, chaude et trouble',[4] that it gives rise to a 'liquéfaction urinaire du ciel':[5] the sky 'pisses' light. The final image of the *récit*, where the narrator sees, in Simone's vagina, 'l'œil bleu pâle de Marcelle qui me regardait en pleurant des larmes d'urine',[6] achieves the intersection of the register of vision and that of liquidity, as the *eye pisses* and *looks* at the same time. This further level of complexity in the associated images and objects of the text renders the structure of these images not simply permutative, in Barthes's sense. While introducing an unrecoverable *vertige* into the text, liquefaction functions on a phenomenological and metatextual level: it signals the liquefaction of reality—its transformation into a flux, a flu-

[3] Ibid., 32.
[4] Ibid., 52.
[5] Ibid., 57.
[6] Ibid., 69.

idity—and the breakdown of the solid outlines of the text, the fluidity of sense transgressing the boundaries of words. Beyond the primary narrative, the one which Barthes identifies as the 'sphère métaphorique pleinement constituée', the provisional closure of the text is achieved by the intersection of the signifiers of *seeing* and *urinating*. This is to be linked to the 'reminiscence', in the second part of *Histoire de l'œil*, by the author of the traumatic origin of the *récit* in the image of the white, upturned eyes of his blind father urinating, and laughing at the same time. I look at the question of this 'origin' on p. 169. What seems evident is that the 'double' chain of roundness and liquidity cannot, on its own, account for the permutations and intersections of the text, which have to be completed by the register of light, seeing, and not seeing.

16. / punning

The contagion which threatens narrative and induces anxiety is not only that between objects, is not limited to the phenomenal domain but takes place also at the level of the signifier, the letter. We have seen already how the story of the eye is placed under the sign of the pun, *auch*, and how it opens with a deliberate linguistic association of heterogeneous, distinct elements which is acted out: 'Les assiettes, c'est fait pour s'asseoir.' The theory of the pun, the *Witz*, proposes that it functions to force together, to juxtapose or superimpose, to bring into proximity objects, elements or ideas which, normally, are separate and distinct. And it is more the *formal* (structural) strategy of this juxtaposition which produces laughter than any qualities inherent in the objects thus brought into proximity. Laughter, or anxiety, both are linked to the *éclat*, the rupture, or, in Barthes's language, *jouissance*. Thus the formal strategy of the association of objects is doubled by an explicit or implicit effect of punning at the linguistic level: 'Les assiettes, c'est fait pour s'asseoir'. Later, Simone will confess her word associations (the text mimics a psychoanalytic session) in response to the words *œil* and *urine*:

> Et comme je lui demandais à quoi lui faisait penser le mot uriner, elle me répondit *buriner*, les yeux, avec un rasoir, quelque chose de rouge, le soleil. Et l'œuf? Un œil de veau . . .[1]

This remarkable series of associations is engaged through the literal slippage from *uriner* to *buriner*, not only associating the semantics of liquefaction with that of light and vision ('le soleil') and thus indirectly determining the 'liquéfaction urinaire du ciel', and with the theme of colour (red as the colour of the blood which 'rises to the head' in moments of sexual excitement or shame), but, at the same time, intertextually signaling Dalí and Buñuel's contemporary image of the eye slit by the razor, the eye 'engraved', 'carved up' ('buriné'). Similarly, the association of eye and egg is overdetermined; it is produced by the association of roundness and whiteness, but also by the linguistic proximity anchored in the syllable *œ: œil–œuf*. This association is explicitly produced by Simone, in a manner similar to Leiris's *gloses* in his *Glossaire j'y serre mes gloses*:[2] œuf = œil de veau. We might, as good psychoanalysts, read these puns as symptoms of the anxiety which inhabits the two characters. The puns may function as a momentary overcoming of the protective shield constructed against the threat of the single, total glimpse of total contiguity, against that trauma. But there is also a level of implicit linguistic slippage in *Histoire de l'œil*, not ascribable to the conscious or unconscious will of the characters or the narrator and thus to an individualized unconscious, ascribable only to the formal strategy of the text. This establishes a further proximity between *Histoire de l'œil* and the writing of Roussel and Leiris, the short-circuiting of linguistic discontinuity and

---

[1] Ibid., 38.
[2] Michel Leiris, *Glossaire j'y serre mes gloses* (Paris: Gallimard, 1939).

distinctness. It makes the reading of *Histoire de l'œil* a radically unsettling experience of the suspicion that *something else* is going on other than the *story* of two apparently disturbed characters acting out their anxieties and desires.

Linguistic slippage functions via the syllable, the 'vocable' for example in this arbitrarily chosen series of words which scan the text: œuf–œil–soleil–couille–cul–cuvette–assiette. It sometimes functions in a different manner, producing near-reversible phrases, such as the chapter titles: *L'œil* de *chat* / Un *tache* de *soleil*. The narrator confesses that he is haunted by names: 'J'acceptai la hantise des noms: *Simone, Marcelle*'.[3] This haunting by the name suggests the importance of the level of the sonorous signifier in the text, the *name*, as a signifier that resists exchange, translation into meaning conceived as communication. That the narrator's discourse, the *récit* as such, is *haunted by the name*, underlines the importance, which is nevertheless not reducible to a structural permutation, of the materiality of words and their proximity on this level.

Words, like objects, in normal, worldly, discourse, are distinct, organized in sentences according to a hierarchy, syntax. But if the distinctness and discontinuity of objects, in a single, terrifying glance, is seen as a virulent molecular flux, as a formless totality of matter, or, in other words, as nothing, words too can be revealed, in a single glance, as simply discontinuous fragments of pure unmediated sound, line or sense. We never go that far, there is only the suspicion, through punning and linguistic short-circuiting of the kind described, of this formlessness, this lack of figure/total figure. The glimpse that would reveal the totality of traces left in the labyrinth of thought always hinges back into discontinuous, linear discourse: 'tout homme est suspendu aux récits', suspended by the hook of the *récit*, on the edge of the maelstrom.

[3] *O.C.* I, 24.

17. / œ II

The **o** figures the hole, the maelstrom, the circle that may be neither ground nor figure, the pupil of the transfixed eye. It is also a figure of closure. The **e** is the **o** ruptured, the **o** slit, mutilated. On its own it is a figure of the transgressive opening of closure and an anti-retinal image close to the razored eye of *Un chien andalou*; it is a figure of ruptured closure and of violence done to the eye. As **œ**, the letters, forced together, represent the suspension, the hinge or hook that locates us within discourse. It prevents us from falling into the abyss of the maelstrom, the formless void of the hole, and positions us in discourse and in language.

18. / eye as figure

The eye is an iris surrounded by a pupil in an ellipsoid 'white', sunk into a socket in the skull. Its interior structure resembles the target. It is, at least, a hole in a space. Or is it a spot against a background? The structure of the target makes it impossible to decide which is ground and which is figure. Both ground and figure. Total ground/total figure. The eye as figure, the figure of the eye, produces a disturbing *vertige*, an inability to be located, to locate oneself.

This indeterminacy characterizes a series of figures in the text, in which circular or round figures are outlined, their outlines then problematized through either a kind of reversal, or an implosive collapse. Thus, when the narrator and Simone visit the asylum the first time, there occurs a scene whose details have an extremely oneiric quality. As in the image of the slit eye in *Un chien andalou*, the moon ('disque lunaire'[1]

---

[1] *O.C.* I, 580 (i.e. in the second version of the text, Bataille's addition).

resonating with a later description of the 'globe oculaire')[2] cuts through the clouds and illuminates a large stain in the middle of the white sheet which Marcelle has hung out the window, and then is masked again. The moon—white disc against the black night—is reversed by the image of the dark stain against the white sheet. The narrator also mentions, as a 'chose curieuse',[3] that at this scene Simone and Marcelle are wearing *black* and *white* stockings, respectively. A little further on the image of the moon is repeated, but superimposed on that of the image of the bright window in the midst of the darkness: 'trou rectangulaire perçant la nuit opaque'.[4] Both of these images repeat the fundamental figure of the eye, or the target, and other such images are produced in the text—the egg, for example. We are also able to read the incidence of a number of such double figures: at the outset the narrator and Simone often look fixedly at each other, their eyes locked into a fixed mutual stare. The text will operate according to displacements of the eye as the object of vision—the narrator or Simone subsequently stare fixedly at eggs ('fixer sur les *œufs* les *yeux* grands ouverts'),[5] at the 'trou éclairé de la fenêtre vide',[6] until finally it is again the *eye* which looks back at the eye: 'je *vis* exactement, dans le vagin velu de Simone, l'œil bleu pâle de *Marcelle* qui me regardait . . .'[7] The fundamental figure of the text, then, is that of the eye in a fixed stare, a mutually fixating

---

[2] Ibid., 54, 75.

[3] Ibid., 31. The phrase 'chose curieuse' appears a number of times in the text, announcing the juxtaposition or superimposition of objects. Cf. *O.C.I*, 21: 'Chose étonnante, cela me redonna du cœur au ventre. On allait accourir, c'était inévitable'; 67: 'Chose curieuse, nous n'avions aucune préoccupation de ce qui aurait pu arriver'; 74: 'Chose curieuse: je ne fis aucun rapprochement entre les deux épisodes avant d'avoir décrit avec précision la blessure faite à Manuel Granero . . .'. The phrase seems to suggest the lack of intentionality or cynicism which informs the eroticism both of the characters within the text and the author, who are *subjected*, as almost innocent victims, to the bizarre permuations which the text operates.

[4] *O.C.* I, 31.

[5] Ibid., 37.

[6] Ibid., 32.

[7] Ibid., 69.

gaze which radically reduces the sight of anything else, the usual unfolding of the field of vision and the location of the subject at the origin of a visual field. The text tells a *story* which moves from one fixed mutual gaze to another, the movement being that of the location of the eye. The very movement of figure, indeed, is commented on by Derrida in 'La mythologie blanche' as having its 'origin' in the act of turning one's eyes from the sun—the turn of the trope is the turn away from the blinding vision of the sun.[8] In Bataille's text too, the operation of figure, the movement of displacement from one image to another, may be read as a movement away from the Icarian impossibility of staring at the sun, the fixed and blinding gaze. The mutual stare repeats this tautological immobility.

The fact that the text pays attention to such double figures and to the reversals that are operated suggests the extent to which the text functions not according to linear narrative progression, but according to structural, spatial movements such as reversal, displacement and superimposition. The figure of the eye, as Foucault notes in *Préface à la transgression*, is a privileged one. It is linked to the instant, the blink of the eye, the moment of rupture. It is that moment of undecideability between figure and ground that the eye as figure highlights. The fixed stare radically annihilates the mobility of the eye as the centre of the visual field, locks it into fixity and thus into a figure.

19. / *rose et noire*

The extant English translation of *Histoire de l'œil*, published by Penguin, features on its cover an image from Marcel Duch-

---

[8] Derrida, 'La mythologie blanche', in *Marges de la philosophie* (Paris: Minuit, 1972), 289–307.

amp's painting *Etant donné le gaz d'éclairage et la chute d'eau,* a usage which seems quite arbitrary.[1] But how might Duchamp be related to Bataille? The question can lead us into an intertextual reading of *Histoire de l'œil.* As a radical, transformative text, in a number of instances it takes 'ready-made' signifiers from other sites, and weaves them into the texture of its own narrative, much in the same way as Duchamp employed his 'ready-mades'. In effect, the intertextuality of *Histoire de l'œil* is fairly explicitly signaled in the text if one distinguishes a deliberate, strategic intertextuality from a more general and implicit level of intertextual relation which would read the book, for example, against Apollinaire or Roussel. The explicit and strategic functioning of intertextuality uses what I have called 'ready-made' signifiers signaled through either italics, quotation marks or proper names ('bonnet phrygien',[2] Don Juan,[3] 'rose et noir'[4]), which represent in the text a whole

---

[1] This use of Duchamp's image is not justified through any proximity between Bataille and Duchamp. Duchamp is hardly present in any of the contexts with which Bataille is involved. It is undoubtedly the result of a desire to associate the book with the connotations of the image, which shows a featureless or formless naked body cut off at the shoulders. The image would thus associate Bataille with a generalised 'Surrealism' and signal the sexual explicitness of the text. Originally, however, this image would have been seen through a keyhole positioned at chest height, so the viewer would have to bend down to look through it and see the image. Taking the image out of this apparatus erases the experience of vision Duchamp had wished to construct; the difficulty of seeing, the fact of it being only *you* seeing, the physical experience of seeing this image are part of Duchamp's *anti-retinal* 'aesthetics' (cf. Krauss, *The Optical Unconscious*, 95–142), a desublimatory movement whereby the experience of seeing becomes a bodily one. The viewer is thrown back upon themselves as voyeur, as sexualized, desirous body. Through circumstance, then, the association with Duchamp points to the same process operative in Bataille's text: a line from the text reconnects the eye to the body and plugs the body into the network, rendering it unstable and fragmentary.

[2] *O.C.* I, 43: 'je me rappelai alors la peur affreuse que j'avais faite à Marcelle quand elle était sortie de l'armoire et en particulier deux détails atroces: j'avais gardé sur la tête un bonnet Phrygien, accessoire de cotillon d'un rouge aveuglant . . .'.

[3] Ibid., 58 ff.

[4] Ibid., 13, 56.

anterior textual body. It is the transformation and use of this body which I now consider.

> Il faisait extrêmement chaud. Simone plaça l'assiette sur un petit banc, s'installa devant moi et, ne quittant pas mes yeux, s'assit sans que je pusse la voir sous sa jupe tremper ses fesses brûlantes dans le lait frais. Je restai quelque temps devant elle, immobile, le sang à la tête et tremblant pendant qu'elle regardait ma verge raide tendre ma culotte. Alors je me couchai à ses pieds sans qu'elle bougeât et, pour la première fois, je vis sa chair 'rose et noire' qui se rafraîchissait dans le lait blanc. Nous restâmes longtemps sans bouger aussi bouleversés l'un que l'autre . . .[5]

Italicization and citation marks can function in a text to indicate either a name, a particularly important signifier, or a piece of text imported from elsewhere, whether that other place is another text or reported speech. The names *Simone* and *Marcelle*, as we have seen, are italicized in a section of the text where the narrator speculates on the obsessive power these words have for him, their *hantise*. The 'being haunted' of the text is thus signaled in the above quotation by the quote marks around 'rose et noire'. They alert us to a haunting, to the suspicion of another body inhabiting, spectrally, the space of this text.

An initial intertext can come from the future of Bataille's writing. 'Le rose et le noir' in Bataille's universe is part of a dynamics of the erotic. In his essay on Proust in *La littérature et le mal*, Bataille proposes: 'Si l'amour est parfois rose, le rose s'accorde avec le noir, sans lequel il serait le signe de l'insipide. Sans le noir, le rose aurait-il la valeur qui atteint la sensibilité? Sans le malheur à lui lié comme l'ombre à la lumière, une prompte indifférence répondrait au bonheur.'[6] 'Rose et noir' thus functions as a signal of the transgressive erotics of the text which will always associate the pure with the impure, show the complicity of any image with its reversal, open out any signifier to an *informe* oscillation.

[5] Ibid., 14.
[6] *O.C.* IX, 267.

The couple 'rose et noir' thus also signals a *political* (polemical) opposition between Breton and Bataille around 'la rose' as flower. In *Nadja*, published in 1928, the eponymous heroine proposes: 'Le *rose* est mieux que le *noir*, mais les deux s'accordent.'[7] For Bataille, however, there is no 'accord', no synthesis in a reality transcendent of the opposition of the colours, or of the necessary distinction between the pure and the impure. Breton, in the 'Deuxième manifeste du surréalisme', takes issue in particular with Bataille's article 'Le langage des fleurs',[8] where Bataille had proposed that *the* rose is the site of a contradiction between base materiality and spiritual beauty. For Bataille the rose arises from 'la puanteur du fumier'[9] and withers away, after its death, into putrefaction. He adds that: 'l'intérieur d'une rose ne répond pas du tout à sa beauté extérieure, que si l'on arrache jusqu'au dernier ses pétales de la corolle, il ne reste plus qu'un touffe d'aspect sordide'.[10] This gives rise to the sentence, or rather, the 'ink-stain', as Bataille puts it, '*l'amour a l'odeur de la mort*'.[11] The beauty of the rose, then, elicits desire as the site of an impossible contradiction between base materiality and spiritual ascension, between 'le noir' and 'le rose'. For Bataille, even before being coupled with 'le noir' or 'noir', *the* rose is an object of profanation, a truth affirmed by the anecdote of Sade's ordering roses in the Bastille only to pluck off their petals and throw them into a stinking pile of liquid manure.[12] Breton rejects Bataille's affirmation of base materiality as obsessive, and describes the rose as a paradigmatic figure of itself: 'Il n'en reste pas moins vrai que la rose, privée de ses pétales,

---

[7] André Breton, *Nadja* (Paris: Gallimard, 1928), 138.
[8] Breton, 'Deuxième manifeste du surréalisme', in *Manifestes du surréalisme* (Paris: Gallimard, Coll. Idées), 144–9.
[9] Bataille, 'Le langage des fleurs', *O.C.* I, 176.
[10] Ibid.
[11] Ibid.
[12] Ibid., 178: 'Et le geste confondant du Marquis de Sade enfermée avec les fous, qui se faisait porter les plus belles roses pour en effeuiller les pétales sur le purin d'une fosse, ne recevrait-il pas [. . .] un portée accablante'.

reste *la rose . . .*'.[13] He reads Sade's gesture as determined by a desire to rid the rose of its stereotypical poetic associations, to enable mankind to rid itself of its chains, to reduce the rose to its essence. Breton's gesture is essentialist and tautological (the rose *is* the rose). It is countered by Bataille's transgressive reading of 'le rose' with 'le noir', *at the same time*. The pink, or the fleshy, *and* the black, or dark.

Breton's definite articles (*le* rose, *le* noir) transform the colours from their status as signifiers into archetypes, symbols or objects. Bataille's proposition of 'la chair rose et noire' resists this symbolization, refuses the transposition of the play of colours and highlights their transgressive interplay. 'Rose et noir' do not *stand for* archetypal essences, but are caught in the informal oscillation of their qualities. Denis Hollier writes that 'nothing is more excluded by Bataille's texts than that eroticism be fetishised'.[14] The counterpart of the mutilating text, severing body parts from the unification of the body, from the transcendental signified of 'the body', is a refusal of fetishism, since fetishism, in the Freudian schema, points to the fetish as *standing for* something else, in effect, the mother's castrated penis. In other words, this is not a text about castration. Bataille's writing is an attempt at anti-poetic, non-metaphoric fetishism, in which the living organ *is* the living organ, but simultaneously the dead thing. 'Chair rose et noire' does not *stand* for anything, it *is* Simone's flesh: the copula *is* again suggesting a difference rather than essence.

In relation to this polemical opposition, we can read 'rose et noir' not as metaphor but as the sign of an undecideable oscillation, and this undecideable opposition also characterizes the intertextual relation which the phrase sets up with other texts. 'Rose et noir' does not refer to its intertexts, but provokes an infectious ruining of them in relation to itself.

The phrase 'rose et noir' appears in two separate poems by Baudelaire. We should note here that *Les fleurs du mal* and

[13] Breton, 'Deuxième manifeste du Surréalisme', 148.
[14] Hollier, *La prise de la Concorde*.

the *Petits poèmes en prose* are in fact privileged pretexts for Bataille's *récit*. The phrase *L'œil de chat*, which titles one of the chapters of *Histoire de l'œil*, appears, in the *Spleen de Paris*, in the poem 'L'horloge', as a way of telling the time ('Les Chinois voient l'heure dans l'œil des chats'[15]) transformed into a lack of temporality ('une heure immobile qui n'est pas marquée sur des horloges')[16] which seems to echo the non-narrative, atemporal and instantaneous *cut* of Bataille's text. The story of the painter's apprentice who hangs himself in a wardrobe, in Baudelaire's prose poem 'La corde',[17] echoes the fate of Marcelle, who hangs herself in an 'armoire normande'. The source of this story (the painter himself) is moreover none other than Edouard Manet, bringing us back by a circuitous route to 'rose et noir'. Manet's painting *Lola de Valence* is the immediate referent for Baudelaire's verse poem of the same name, which includes the phrase 'rose et noir'. In this circular itinerary from Bataille to Baudelaire to Manet, Manet appears as a point of juncture, an intersection or perhaps a destination; the reading of relations along this axis (Bataille–Baudelaire–Manet) is a risk in which my reading of the text engages. This risk involves the body and the gaze, for Manet's paintings set up a desiring gaze and a resisting body, a resisting gaze and a desiring body, which opens and extends the reading of Bataille.

Baudelaire's two poems read as follows:

> *Lola de Valence*
> Entre tant de beautés que partout on peut voir
> Je comprends bien, amis, que le désir balance,
> Mais on voit scintiller en Lola de Valence
> Le charme inattendu d'un bijou rose et noir.[18]

---

[15] Charles Baudelaire, *Spleen de Paris: petits poèmes en prose* (Paris: Garnier Flammarion, 1967), 75.
[16] Ibid.
[17] Ibid., 115–18.
[18] Charles Baudelaire, *Les fleurs du mal* (Paris: Garnier Flammarion, 1964), 181.

Toute entière

. . .
Parmi toutes les belles choses
Dont est fait son enchantement
Parmi les objets roses et noirs
Qui composent son corps charmant,

Quel est le plus doux? . . .[19]

Both poems are readable as a discourse on textuality, or on
meaning as such, a discussion which takes place, significantly,
between men only. They foreground the *impossibility of
deciding* in a masculinist universe where desire and meaning
are ruled by a more or less virile linearity. The texts undermine
this linearity and emphasize the undecideability of the destination of desire. Which beauty, which part of the body? *Lola
de Valence* suggests that while male desire and the male organ
swing from one object to another, at least Lola (a pictural
representation rather than a writerly one) and her sex ('bijou')
can be relied upon to capture and fix this desire. But it is 'rose
et noir'. Not one or the other but both at the same time. In
'Toute entière' the poet cannot decide between the *objects*,
'roses et noirs', of his lover's body (another mutilation and
defiguration of the unified body). The phrase 'rose et noir' is
at the nexus of an oscillation, an undecideability, an impossible
contradiction.

Looking at the painting of Lola, a fully clothed Spanish
dancer, the 'rose et noir' appears to refer to her face fringed
with dark hair. Susan Suleiman comments on the 'displacement downwards' operative in Baudelaire's text, from the face
to the genitals, pink flesh fringed with dark pubic hair,
although she does not refer to the painting.[20] Bataille makes a
similar remark in *L'érotisme*. Referring to Leonardo da Vinci's

---

[19] Ibid., 67.
[20] Susan Suleiman, 'Transgression and the avant-garde', in *Subversive Intent*
(Cambridge, Mass.: Harvard University Press, 1990), 81. Suleiman links the
image to the painting *The Rape* by René Magritte which forces the tenor
and vehicle of Baudelaire's metaphor together, making the functioning of the
metaphor explicit.

notebooks, he counters the assertion that the ugliness of the sexual act is excused by the beauty of the face and dress of the participants: 'Léonard ne voit pas que l'attrait d'un beau visage ou d'un beau vêtement joue dans la mesure où ce beau visage annonce ce que le vêtement dissimule. Ce dont il s'agit est de profaner ce visage, sa beauté.'[21] The beauty of the face cannot be considered separately from the 'ugliness' of the sexual parts; it is locked into a dialectic of transgressive profanation whereby the function of the beautiful face is paradoxically to point to the hidden sex. Lola's face, 'rose et noir', read or seen through Baudelaire's phrase 'bijou rose et noir', points downwards to her hidden sex.

What happens in Bataille's text, and what different reading does it imply of Baudelaire's poems? If, as I have suggested, Bataille's text refuses transposition, makes the sex visible *as an organ*, it refuses the structure through which the female sex is the destination of all eroticized metaphors. It refuses and deconstructs the mechanism whereby the female sex is the origin and limit of erotic meaning. The sex is seen *as* sex, not as hidden, secret ontological source. It may, as Suleiman suggests, operate a 'displacement upwards', from the sex to the face, reversing the direction of Baudelaire's metaphor, but the relation here is more one of displacement or infection than of transposition, of a mimetic *standing for*. Bataille's text, then, problematizes the metaphoric relation of Baudelaire's poem to Manet's painting *Lola de Valence*, diverts the signification of the poem away from metaphoric transposition.

The undecideability present in Baudelaire's poems becomes in Bataille's text a contradiction between 'rose' and 'noir'; their impossible but mutual forcing together. This in turn may point to a characteristic of Bataille's text. For, if the narrator of *Histoire de l'œil* and Simone agree that the best name to use for the vagina is 'cul', this suggests a kind of ambiguity, for 'cul' can mean at the same time vagina and anus. When the narrator lifts up Simone's dress to look at her 'parts' he sees

---

[21] Bataille, *L'érotisme* (Paris: Minuit, 1957), 161.

the 'rose' (her vagina) and the 'noir' (the anus), while both parts are forced together in the word 'cul'. What is important in Bataille's textual eroticism is not the predominance of sex, but the permutation of a displacement between organs, and as such, this oscillation or ambiguity shifts the emphasis away from that 'origin'.

Metaphoric substitution is ruined by informal oscillation, and this has the effect of imposing a different mode of reading, one which focuses on the structural interplay of signifiers rather than their mimetic relation to something exterior to the text. To move again into the future of Bataille's œuvre, we can see this mode of reading operating in his account of Manet, and propose the painting *Olympia* as a hidden intertext for 'rose et noir'. In his 1955 text *Manet*,[22] repeating a gesture made by Paul Valéry, he displaces the referent of Baudelaire's poem from the painting *Lola de Valence* to *Olympia*.[23] Biography can inform us here. It is reported that Bataille, on visiting the Louvre at the time of the composition of *Histoire de l'œil*, was *transfixed* in front of this painting, and that among his loans from the BN at the time of writing *Histoire de l'œil* were several books on Manet. As a reader of *Histoire de l'œil*, I am sent in front of this painting.

Before indulging in the pleasure of looking, it is worth exploring the way in which Bataille's analysis of Manet's painting proposes it, implicitly, as functioning in a similar way to the text of *Histoire de l'œil*. In his 1955 book on Manet, Bataille relegates the painting *Lola de Valence* (1862) to Manet's Spanish period, which he characterizes as 'only a passing phase, not to say a dead end', pausing to mention Baudelaire's role in persuading Manet to 'go Spanish'.[24] Bataille presents his vision of Manet's role as inaugurator of modern painting via Malraux's discussion of Goya, proposing that, while Goya is modern through excess, through going one

[22] *O.C.* IX.
[23] Ibid., 141.
[24] Ibid., 122.

better than 'les solonnelles *conventions* du présent',[25] Manet's
painting responds through absence. It is a question, Bataille
proposes, of wresting objects from their debased function as
part of a utilitarian tradition, from a 'bourgeois torpor', of
attaining 'quelque réalité indiscutable, dont la souveraineté
ne pût être pliée par un mensonge à l'immense machine utili-
taire.'[26] Manet achieves this through silence. Significantly,
given Hollier's postulation that all of Bataille's work is an
attempt to destroy the cathedral that his early text 'Notre
Dame de Reims' constructs,[27] Bataille refers to Malraux's
suggestion that 'la peinture moderne dans nos musées est
notre seule cathédrale édifiée par le temps présent'.[28] But,
Bataille adds, it is essentially a secret cathedral. The modern
world cannot raise the sacred into a majestic form, can only
experience 'une transfiguration intérieure' which is 'en quelque
sorte négative'.[29] The sacred is mute, and painting opens into
silence. This silence is achieved through a 'destruction du
sujet',[30] an indifference to the subject of the painting, which
serves only as a pretext for the painting.

Thus the *painting* of Manet 'obliterates' the *text*, or whatever intentional meaning resided in the subject: 'ce que le
tableau signifie n'est pas le texte, mais l'effacement'.[31] The
destruction of the subject, the obliteration of the text correspond to the refusal of transposition in Bataille's writing, and
to the destruction of the interior eye behind its surface, the
emasculation of phenomenological seeing. Organs are not
fetishes, metaphors for a transcendent origin or destination,
but are revealed in their physical presence, as objects. Manet
offered up the starkness of 'ce qu'il voyait'.[32] That is, he offered

[25] Ibid., 134.
[26] Ibid., 135.
[27] Hollier, *La prise de la Concorde*, 32: 'Bataille n'écrira que pour ruiner cette
cathédrale.'
[28] *O.C.* IX, 135.
[29] Ibid.
[30] Ibid., 125.
[31] Ibid., 142. In italics in the text.
[32] Ibid. In italics in the text.

only raw vision, without thought. Painting, and writing, mutilate the subject from the majestic form which was its unification, exposing the body of that subject (painting or text) to a transgressive play that opens into silence, into absence, or, in terms more proximate to our own, to the *informe*, to *vertige*. Words are not subjected to the meaning that lies behind them, but imposed in their materiality, which causes a play of displacement similar to the play of colours in painting. The meaning of the painting is obliterated, imposing the painting as a play of colours, in the same way that the meaning of Bataille's *histoire*, the fiction, is subordinate to the play of signifiers across the text: '*Olympia* parvint à la raideur, à la matité de la violence: cette figure claire, composant avec le drap blanc son éclat aigre, n'est atténué par rien. La servante noire entrée dans l'ombre est réduite à l'aigreur rose et légère de la robe, le chat noir est la profondeur de l'ombre . . .'[33] The play of 'rose' and 'noir' in Manet's painting corresponds to the textual displacement operative in *Histoire de l'œil*, so that the phrase 'chair "rose et noire"' in the text, through its intertextual resonances, functions as a metatextual pronouncement; it tells us, through Manet, how the text works.

Back in front of the painting, there, as readers or spectators, we are caught between two gazes, that of the model Victorine Meunier, and that of the cat, *l'œil de chat*. Tracing the intertextual itinerary Bataille–Baudelaire–Manet has led us to a place where we are under the gaze of the cat, implying, through the chapter title of *Histoire de l'œil*, that the position of the reader of the text and of the spectator in front of the painting coincide. That this is the place (at least temporarily) from which to read the text. Of course, following Bataille's text's refusal of metaphor and its diversion of Baudelaire's 'rose et noir' away from Lola, we are looking for the 'chair rose et noire' of the female sex. We do not get it. Victorine's hand lies splayed across it, in a gesture which simultaneously hides and reveals it. For, if her covering of her sex draws attention

[33] Ibid., 147.

to it through her sly prudishness ('Je pense comme une fille enlève sa robe' writes Bataille), it dispenses with the necessity to refer to it through metaphoric transposition. Victorine is pointing to her 'chair rose et noire', and the black cat is absolutely not (if we read the painting through *Histoire de l'œil*) a metaphor for her 'pussy'. On the one hand it is a 'chat' and not a 'chatte' (its upright tail suggests as much), and on the other hand its function is determined by the chapter title 'l'œil de chat'. The covering/pointing gesture is a refusal of fetishism; it destroys the 'subject' of sex and liberates the painting to the erotic play of colours: the black ribbon around the neck of the model,[34] the rose in her hair, the pink dress of the black servant, the 'coal black' cat, the bouquet of roses. In the same way Bataille's refusal of metaphoric transposition liberates the play of signifiers in *Histoire de l'œil*. If the text and the painting are read as *in relation*, or as transformations of each other, even as metatextual commentaries of each other, we can read the 'chair rose et noire' in *Histoire de l'œil* as the hidden but revealed secret of the painting, not a secret in the sense of original source, but as the key to its functioning as a text of mutilation and of anti-poetic fetishism.

20. / the apparatus

Our intertextual reading of 'rose et noir' has taken us on a circuitous itinerary through Breton, Baudelaire, and Manet, to Olympia herself. As proposed above, this trajectory was determined by a sense of being 'haunted' by another body, a suspicion that led to an investigation. As readers we encounter such an anxiety, such a suspicion, on numerous occasions with

[34] Like 'rose et noir' the ribbon is a signifier which has explicitly programmed a literary text: Michel Leiris's *Le ruban au cou d'Olympia* (Paris: Gallimard, 1981).

Histoire de l'œil, but perhaps nowhere more powerfully than with the figure of Marcelle, whose death itself is a result of a haunting of sorts. The asylum in which Marcelle is incarcerated, after the episode of the 'armoire normande'[1] in which she shuts herself during the orgy, is transformed in the narrator's imagination into a 'chateau hanté'.[2] Again, this transposition is ambiguously both determined and reflected by a linguistic slippage, from 'maison de santé' to 'chateau hanté'.[3] Such a displacement in itself functions intertextually, signaling the indebtedness, the 'hauntedness' of this text by the literary tradition of the Gothic novel, the 'font' of much 'damned' textuality, from Sade and Poe to Lautréamont. Walpole's *Castle of Otranto* is an obvious pretext for this castle. That the asylum also turns out to be a haunted castle indirectly suggests the proximity of the textuality and eroticism of Sade (the 'chateau de Silling' of the *120 journées de Sodom*)[4] and to the discourses of 'madness', both 'removed' from the space of the 'normal', from society. In Bataille's time the notion of 'evil' which informs the Gothic narrative has become that of madness: the haunted castle has become the asylum.

For Marcelle, and for the reader, the haunting does not stop there, however. Linked to the literary history of the asylum/ haunted house and thus to the genealogy of *terror*, Marcelle's imagination brings together a number of intertextual seams: she fears the return of the Cardinal (this section in italics in the text),[5] who is also the *'curé de la guillotine'*.[6] The narrator explains this displacement himself, tracing it to the original trauma of Marcelle's vision of him covered with blood and wearing on his head a 'bonnet phrygien'.[7] The Phrygian

[1] *O.C.* I, 18–22.
[2] Ibid., 41.
[3] Ibid., 42.
[4] Sade's castle also appears at the end of Dalí and Buñuel's *L'âge d'or*.
[5] *O.C.* I, 43: 'Je la regardais avec inquiétude et comme j'avais déjà à cette époque un visage dur et sombre, je lui fis peur moi-même et presque au même instant elle me demanda de la protéger *quand le Cardinal reviendrait.*'
[6] Ibid.
[7] Ibid.

bonnet[8] of the revolutionary, associated thereby with the terror of the guillotine, is confused with the fear and piety in relation to the priest: 'une étrange coïncidence de piété et d'horreur des prêtres expliquait cette confusion'.[9] But the confusion is not thereby explained; there is no necessary link between the revolutionary executioner and the priest. The narrator's explicit explanation has to be accompanied by an analysis not of the psychology of the characters but of the implicit textual and intertextual mechanisms of the text. This will lead us to a further set of associations. The 'armoire normande', which gives its name to the second chapter of *Histoire de l'œil*, is initially used by Marcelle to hide her shame from the other participants in the adolescent orgy: 'Elle voulait se branler dans cette armoire et suppliait qu'on la laissât tranquille.'[10] It is from the 'armoire' that the 'filet' of urine appears when Marcelle urinates. Unable to liberate herself because of her shame, for Marcelle the 'armoire normande' has now become both a 'pissotière de fortune' and a 'prison' ('cette pissotière de fortune qui lui servait maintenant de prison').[11] The precedent of this condensation (wardrobe/urinal/prison) invites us to take the rapid step of seeing the wardrobe now as guillotine, the object now taking on Giacometti and Duchamp like character-istics, as a murderous urinal which is also a prison. The itinerary of this transgressive object does not stop there, however; it is picked up much later in the text, in the final scene in the church of Don Juan. The priest is pictured, 'les mains croisées sur le seuil de *l'armoire*, le regard élevé vers un point du plafond . . .'[12] Simone kneeling by the confessional, the priest enters 'sans mot dire dans l'armoire [et] ferma la

---

[8] See Jean-Luc Steinmetz, 'Bataille le mithriaque (sur *Histoire de l'œil*)', *Revue des sciences humaines*, 206 (1987), for a detailed analysis of Bataille's use of the Phrygian bonnet, symbolizing not only, according to Steinmetz, the Revolution and its apparatus of Terror but also the cult of Mithra.

[9] *O.C.* I, 43.

[10] Ibid., 20.

[11] Ibid., 21.

[12] Ibid., 599 (second version: 'armoire' is substituted for 'cabine' in the first).

porte sur lui.'[13] Like Marcelle, the priest shuts himself into the cupboard, and this is the first step in the superimposition of the image and figure of Marcelle on to his, leading to his death (also by asphyxiation—Marcelle hangs herself in the 'armoire') and to the description of his mutilated eye, in Simone's sex, as *that of Marcelle*. The superimposition of guillotine/urinal/prison on to confessional box is further exploited by the mention of the *œil de guerite*, the eye of the 'box', this 'eye' being exactly superimposable on to the 'eye' of the guillotine, and even the 'hole' of the urinal. The 'filet d'urine' which emerges from the wardrobe as *pissotière* is superimposed on to the stream of blood from the guillotine. Marcelle's association of priest and executioner is thus explained by an association realized further on in the text—she is, so to speak, haunted by the future, or in other words, haunted by the non-linear temporality of the text. Again, the psychological explanation is undercut by textual associations which are spatial and structural, reducing time to a single instant, in which all things are linked.

The figures of the Phrygian cap and the guillotine-like apparatus establish a further intertextual seam for the text. The guillotine, as we know, was invented by the French physician Joseph Ignace Guillotin and proposed to the Revolutionary Constituent Assembly in 1791, during the Terror.[14] One of its notorious locations is on the Place de la Concorde, in Paris, on the site of which now stands the obelisk of Luxor, a monument crucial for Bataille.[15] Historically, the figure of the guillotine relates the text to that moment when the Enlightenment discourses of rationality, divested of the space of the sacred, erupt in a violence produced, one might speculate, by that very divestment, the moment in which that

[13] Ibid. (second version 'armoire' is substituted for 'tabernacle' in the first).
[14] For an analysis of the symbolism of the guillotine in relation to the thematics of torture in modern literature, see Laurent Dispot, *La machine à terreur* (Paris: Grasset, 1978).
[15] Cf. Bataille, 'L'obelisque', in *O.C.* I, 501–13. See also Hollier, *La prise de la Concorde*, the title of which exploits this thematics.

literature of 'terror' which emerged in the eighteenth century is acted out on the political stage.[16] The guillotine decapitates the head of reason; its historical moment is one at which the body takes its revenge upon the head. The urine which seeps from the *armoire* has as ghostly historical after-image the blood which results from this violent overthrow.

The figure of the guillotine is also part of the genealogy of the discourse of the museum, which, as I have suggested earlier, *Histoire de l'œil* subverts and undermines. If sacrifice, for Bataille, is a spectacle which is a necessary element of a relation to the sacred, the geometric and scientific efficiency of the guillotine radically reduces the spectacular nature of the death. The moment of the sacrifice, the moment at which discontinuity passes into continuity, is reduced to a 'clin d'œil', the flash of an eye. It is essentially not *seen*. Whence the necessity of accumulating around the execution by guillotine the accoutrements of theatre. The invisibility of the moment of sacrifice is supplemented by the theatricalization of the execution. Anticipating, as Francis Warin suggests,[17] the photographic shutter, the guillotine execution, through scientific efficiency, renders invisible the death itself, and accumulates a spectacular discourse as supplement, hiding, but premised on, this moment. This 'machine de terreur' is thus read as a condition of possibility for discourses and systems of knowledge constructed around an invisible experi-

[16] Mario Praz, in *The Romantic Agony* (Oxford: Oxford University Press, 1970), in the chapter titled '*le vice anglais*' refers to the French myth of the Englishman 'whose greatest pleasure was to attend executions' (438). Although the myth develops from confused reports of an actual occurrence in 1757 at the execution of the would-be assassin Damiens, it is notably developed in the nineteenth century by the Goncourts, and thus feeds into the mythology of terror which would have informed Bataille's aesthetics.

[17] Francis Warin, *Nietzsche et Bataille: la parodie à l'infini* (Paris: PUF, 1994): 'La guillotine remplace ainsi le chevalet du peintre et anticipe l'obturateur et "l'instantané" photographique. On vous "tire le portrait" à l'instant fatidique où le mort saisit le vif et fixe vos traits et les gravures de têtes coupées, brandies par un bras anonyme, achevèrent de diffuser, de rendre public, de pro-stituer de la façon la plus obscène le moment intime et le plus inaliénable de chacun' (196).

/ the apparatus

ence. It is in this context that Bataille's comments in the *Documents* article titled 'Musée' (cited by Warin) can be understood: 'l'origine du musée moderne (le premier fut fondé le 27 juillet 1793) serait donc liée au développement de la guillotine'.[18] In *Histoire de l'œil*'s perverse apparatus we might read the discourse of museography as one which classifies and encloses in boxes (cupboards), and is thus associated with confinement (the apparatus as prison). A Foucauldian genealogy would indicate without doubt the dependence of a discourse of classification, pigeonholing and confinement on the possibility of a panoptic eye,[19] and establish the link between the prison and the museum. But the transformation of the apparatus (cupboard/prison) into a guillotine suggests the more radical dependence of these discourses on the rationalization of death, the transformation of sacrifice into a rational machine. If we read *Histoire de l'œil*, as suggested, as a key moment in the transformation of Bataille's museography, the appearance of the guillotine, particularly here in the context of vision and of the eye, enables the uncovering of the repressed elements of this discourse, and the emergence of a different, anti-museographic and anti-institutional form of knowledge — that which is exemplified in *Documents*.[20] The displacement from the language of the museum (or indeed the library) to that of the spectacle of the guillotine is topographic too, from the Bibliothèque Nationale less than a mile south to the Place de la Concorde.

In addition to the historical aspects of the intertextual function of the guillotine, it is endowed with a certain symbolic value which François Warin explores in his book *Bataille et Nietzsche*. Its symbolic value is determined by the guillotine's *instantaneous* operation, which figures the blink of the eye, itself a figure for the instantaneous *cut* which produces a scis-

[18] *O.C.* I, 239; Warin, *Nietzsche et Bataille*, 196.
[19] Cf. Michel Foucault, *Surveiller et punir* (Paris: Gallimard, 1995).
[20] Rosalind Krauss, and Denis Hollier, in *October*, emphasize this subversive transformation of the museum and use Bataille, in a sense, to propose a radically different version of the discourse of art history, theory and practice.

sion in time. In the preface to the *Histoire de l'œil* from *Le Petit*, Bataille (or Louis Trente) recalls a drawing used for the previous but destroyed text *W.C.* which functions as a 'reminiscence' of *Histoire de l'œil*:

> Un dessin de *W.C.* figurait un œil: celui de l'échafaud. Solitaire, solaire, hérissé de cils, il s'ouvrait dans la lunette de la guillotine. Le nom de la figure était 'l'éternel retour', dont l'horrible machine était le portique. Venant de l'horizon, le chemin de l'éternité passait là.[21]

Another series of transpositions and superimpositions operates here: the image is at once of a scaffold, guillotine, and the eye is at once the sun, and 'bristling with lashes', insect-like, recalling Miró's art. The image draws attention to the parallel between the instantaneous cut of the guillotine, the blink of the eye, and the flash of the sun or of lightning. The reference to Nietzsche's eternal recurrence indicates that what we are concerned with here is that *cut* in time, 'syncope du temps', which is the entrance door, the 'portico' of eternity, or the juncture where time considered in its discontinuous aspect, as *durée*, is ruptured by the time of the instant which is that of continuity. The eye in the guillotine, 'l'œil de guerite' in *Histoire de l'œil*, thus figures that impossible instant of the rupture of a (narrative) temporality of succession by a vision of continuity—the immediate and instantaneous instant. The eye of the body of knowledge, manœuvring the phenomenal world in a temporality of delay and reserve, is *cut* by the severed eye in the guillotine, or in the *fente* of the sex. Moreover, the figure of the eye in the guillotine, the geometry of which is obviously exploited in *Histoire de l'œil*, symbolizes a desire to maintain a vision of the instantaneity of the operation of the machine—to visualize what is removed from vision by efficient science. The eye without the body symbolizes the emasculation of vision from the body of knowledge and the panoptical knowledge of the body which informs modern scientific and social discourses. If the invention of the guillotine, as

[21] *O.C.* III, 59.

Bataille suggests, is contemporaneous with the emergence of modern museographic discourse, to figure an eye in its hole is to render visible that vision which this discourse represses, and at the same time to undercut the notions of panoptical and speculative vision which inform this discourse.

Reading *Histoire de l'œil* intertextually, then, produces a critical commentary on the systems of knowledge of modernity, through this appeal to the history of the late Enlightenment and the Revolution. It also draws us close to that impossible figure underlying the text: the instant—instant of death, instant of the fixed mutual stare and the tautological gaze at the sun—that instant of vision which undercuts any *seeing*, if seeing is considered as knowing.

Bataille's cupboard/urinal/prison/guillotine/confessional box has a distinct relation to Roussel's contraptions, in *Impressions d'Afrique*, for example, produced through the necessity of bringing together certain signifiers, or to Jarry's locomotive in *Le Sûrmâle*, or to Kafka's apparatus of 'In the Penal Colony'.[22] The textual appearance of this kind of 'apparatus' is, in other words, a trope whose fortune it is instructive to trace. As an aesthetic object it recalls, as I have suggested, those objects produced by art in the wake of Surrealism, by Giacometti, Duchamp and others. But in the textual domain, it seems also to have a certain figural role as a representation of the oper-ation of text as such. The apparatus, in this history, is a figure for a series of relations, flows and forces. We might therefore construct a hypothetical literary history of the apparatus as follows:

Mallarmé's poems construct a complex syntax as a network of intersecting lines, establishing each signifier as potentially determined by a multiplicity of others. The poet tries to con-struct a machine to annul chance through a precise spatial arrangement, the guarantee of syntax, the permutative *combi-*

---

[22] Cf. Dispot, *La machine à terreur*, 131–40, and Steinmetz, 'Bataille le mithriaque', 174.

natoire of the dice throw, but *un coup de dés n'abolira jamais le hasard.*

For Raymond Roussel: the apparatus of language becomes almost self-sufficient. The network is turned into a circuit; the end of the text joins up with the beginning, and the procedure attempts to set up perpetual motion, the infinite generation of variants. If you make the apparatus into a circuit, the subject, human agency, becomes redundant, or rather, it becomes the victim of the apparatus which is represented, thematically, as a torture machine, which runs on its own. But, as Kafka's parable teaches, the machine of perpetual motion always breaks down and entails the sacrifice of its engineer: Roussel's suicide in Palermo in 1933 is part of the self-perpetuating machine of his texts and their auto-explication.[23]

Leiris's puns in *Glossaire j'y serre mes gloses* reduce the Rousselian game to a joke, a *Witz*, and despatialize, contract, the Mallarmean topography of chance.

Kafka's State Machine inscribes the Law on to the body of the subject.

There is a sense of the repetition of a primal scene, the torture of a body: dissection, writing, entrapment, a scene which betrays a fundamental unease about the relation of the body to the structure. What happens when you plug the body into the apparatus? It gets divided up into zones of intensity, into organs, it gets written upon, permuted. A writing on or of the body, an engraving (*burin*) is the figuration of the body's consciousness of the apparatus. Bataille's text constructs a Surrealist-type collage or superimposition, a torture machine, which is related to Masson or Giacometti's cannibalistic furniture, but more importantly to a current of imagery in pre-Surrealist texts. ' . . . beau comme la rencontre fortuite sur une table de dissection d'une machine à coudre et d'un parapluie'[24]

---

[23] Cf. Michel Foucault, *Raymond Roussel* (Paris: Gallimard, 1963).
[24] Le Comte de Lautréamont, *Les Chants de Maldoror* (Paris: Gallimard, Coll. NRF Poésies), 234.

/ the apparatus

(*Les Chants de Maldoror* is among Bataille's requests from the BN in 1925). The Surrealist commentary on this infamous phrase might run over and efface the extent to which it presents an image of potential violence, not only in terms of the forcing together of distinct objects, but in terms of the potentially violent qualities of these objects: *dissecting table, sewing machine, umbrella* (with spokes). A superimposition of torture implements similar to the *armoire* collage operates here. *L'anus solaire* also figures, in its representation of a contagion of objects: a 'machine à coudre', a 'parapluie', 'une locomotive composé de roues et de pistons.'[25]

Roussel's textual procedures exerted a fundamental influence on the writing of Leiris, whose texts *Langage tangage*[26] and *Glossaire j'y serre mes gloses* show the workings of a determination of the signified by the signifier. In Roussel's texts, the *récit* is generated by the movement from one reading of a phrase to another, which in his posthumous *Comment j'ai écrit certains de mes livres* he calls 'création imprévue due à des combinaisons phoniques'.[27] Leiris's parodic dictionaries also employ this formula, generating signifiers from signifiers, rather than 'defining' (fixing) them in relation to definitions (signifieds). Allan Stoekl writes in his *Politics, Writing, Mutilation* that: 'both Roussel and Leiris attempt to see the fundamental statements of language as functions of puns: their writing is at least to a certain extent generated out of phonic similarities between words, similarities that might then uncover rather bizarre links between the meaning of words'.[28]

But Roussel's text also feature instruments of torture. Stoekl reads this as the *return* of a violence which had been projected into a realm of textual utopia, language reflecting

[25] *O.C.* I, 82.
[26] Paris: Gallimard, 1985.
[27] Raymond Roussel, *Comment j'ai écrit certains de mes livres* (Paris: 10/18, 1963), 23.
[28] Allan Stoekl, *Politics, Writing, Mutilation: The Cases of Bataille, Leiris, Blanchot, Roussel and Ponge* (Minneapolis Minn.: University of Minnesota Press, 1985), xi.

upon itself, destroying itself, in content, or as an infliction upon the subject. Thus 'The inner violence of major characters (on a thematic level) mirrors the self-directed violence of language (in silence, in wordplay, in *writing*)',[29] and 'these five authors (Roussel, Leiris, Bataille, Blanchot, Ponge) focus on the problem of automutilation and sacrifice'.[30] The ultimate figure of this mirroring may be Roussel's representation (in *Impressions d'Afrique*) of the tattooing on to the skin of a victim of the map of Africa, a torture which is itself determined by a displacement of a previous episode of the text. Such a figure is again represented by Kafka in 'In the Penal Colony', with the 'apparatus' whose function is to write into the victim's skin the 'sentence': 'Honour thy Superiors'. The black humour of the *récit* is rendered grotesque by the pun between 'death sentence' and the sentence written into the man's body. In *Histoire de l'œil* the self-determination of the text by linguistic displacement, on the level both of the signifier (the word) and of the referential qualities of the objects (liquidity, roundness), is an infliction, a structural apparatus, whose violence is mirrored, or is *returned* as the violence to which the protagonists are subjected: Marcelle's almost literal execution through the associations of the *armoire*, the priest's enucleation so that his eye might be brought into line with other signifiers, but also the *inner violence* to which the narrator and Simone are subject owing to the mentioning of certain words and the forging of certain associations.

Torture, of course, is not solely a modernist theme. It is prefigured in Sade, in Poe, and, for example, in the novel *Le Jardin des Supplices* by Octave Mirbeau, all texts which inform the literary imagination of the early twentieth century. It is a particularly privileged theme of the literature of terror, of course, whose pretextual importance for *Histoire de l'œil* is evident through the generic transformation of the Gothic novel, the proximity of the Sadean imagination, and through

[29] Ibid., xii.
[30] Ibid.

the allusion to the Terror. But in the modernist imagination torture is associated with the thematics of the *apparatus*, of which the closest instance, so to speak, is the structure of language. The sense of the determination by the signifier, by the autonomous structure of language, is associated with torture, victimization; subjectivity is associated with being subjected.

21. / out of this world!

The *story* of the eye is not only, in the *récit* and in the second part, 'Coïncidences', the narrative of what happens to it. It is also a speculative reflection on the erotic. The narrator *theorizes* his own experience, much along the same lines that Bataille will later, in texts such as *L'érotisme* (1957). The key to this theory of the erotic is the sense of rupture, of transgression of the limits of *this world*, a sense of a momentary passage, through a vision which is at the same time an emasculation of vision, into *another world*.

> A d'autres l'univers paraît honnête parce que les honnêtes gens ont les yeux chatrés. C'est pourquoi ils craignent l'obscenité.[1]

The castrated eye is the eye of horizontal, phenomenological vision, the eye which receives stimuli which the brain translates into knowledge. It is castrated because it has fallen from an original verticality. The opposing figure is Bataille's *pineal eye*, the eye on a stalk which looks upwards, to the heavens and into the sun. But what the pineal eye sees can only be momentary, instantaneous, because it sucks phenomena into a hole of total continuity. What the pineal eye sees is *nothing*, and is figured by the *Augenblick*, the instantaneous shutter which, in an instant, cuts into time. The blink of the vertical

---

[1] *O.C.* I, 45.

eye is that instant in which the obscene formlessness of the universe is visited upon the subject; vision as the field in which the subject is located in the phenomenal world is torn into with this cut.

Not so much *another world*, then, as a hole or a tear in this world which destructures and radically deforms it. Nevertheless, from within this world, this side, as it were, of the cut, the hole is constructed as a space, which can be occupied, and given a number of different figures, according to the reading one adopts: it is at the same time the realm of the sacred, that of death, and that of madness. What is sure is that the instantaneous glimpse, the passage into this projected other world is induced both by the erotic *vertige* in which the limits and contours of objects are blurred and transgressed, and, in a more obvious sense, by the orgasm. In a social as well as a metaphysical sense, the two main characters of *Histoire de l'œil* are 'out of this world'.

Erotic transgression ruins the stability of the 'normal'. It is in this sense that the cosmic aspect of Bataille's eroticism is linked directly, as part of the same movement, to the obscene and pornographic quality of the text. The *épatement* of the bourgeoisie, the ruining of 'good conscience' and everyday morality are part of a movement away from the world and from the 'worldly'. But the characters of *Histoire de l'œil* are not integrated into any social or anti-social body, like those dystopian communities featured in the works of Sade, the 'amis du crime'. Their anti-social behaviour is not part of any political or social project but simply the first part of a movement away from the limited space of the 'worldly'. The narrator confesses: 'Je n'ai jamais eu en moi la possibilité de prendre ce qu'on appelle une attitude.'[2] The space of the world is that in which objects can be distinguished from one another, can be picked up and used as tools, and where, as Simone suggests, the erotic is limited to sex, 'dans un lit, comme [avec] une mère de

[2] Ibid., 23.

famille'.[3] In tragic space, this departure from the order of the world would lead to punishment and resolution, but *Histoire de l'œil* is no tragedy.

The transgression of the characters of *Histoire de l'œil* is therefore not aimed at the delineation of a formalized space 'beyond', but is something like a hole in the material tissue of 'the world', an interruption which undoes its coherence. The narrator writes 'Cela se passa comme si je n'étais plus sur la terre',[4] the 'comme si' establishing an ambiguous and paradoxical relation to the real. Later, after Marcelle's death, Simone is described as follows:

> Mais il y avait un grand changement dans Simone après le suicide de Marcelle, elle regardait tout le temps dans le vague et on aurait dit qu'elle appartenait à autre chose qu'au monde terrestre où presque tout l'ennuyait; ou si elle était encore liée à ce monde, ce n'était guère que par des orgasmes rares mais incomparablement plus violents qu'auparavant.[5]

Certain aspects of this description, the distant, unfocused gaze and the seeming inhabiting of a different world, establish a proximity with the literary and artistic figure of the female hysteric, much celebrated at the time by the Surrealists. Breton's statement 'La beauté sera CONVULSIVE ou ne sera pas',[6] at the end of the book *Nadja*, published in the same year as *Histoire de l'œil*, signals the importance of this influential thematics within Surrealist iconography. It will also emerge in Breton and Eluard's *L'immaculée conception*, in Buñuel's *L'age d'or*, and most remarkably in Dali's photo-collage *Le phénomène de l'extase*. This celebration of the female hysteric may have taken its direct inspiration from psychoanalytic and psychiatric material, the performances of Charcot, for example, or documents showing the female hysterics at the

---

[3] Ibid., 25.
[4] Ibid., 29.
[5] Ibid., 49.
[6] Breton, *Nadja*, 190.

Salpetrière hospital.[7] In the early '30s it would also take in the figure of the murderous woman—Violette Nozières, or the Papin sisters,[8] and in 1933 the case of Aimée which would be the subject of a lengthy study by Jacques Lacan.[9] The influence of Bataille's heroines (Simone and Madame Edwarda) informs Lacan's well-known take on Bernini's Saint Teresa in *Encore*: that she is coming ('elle jouit')[10]; in a less specific sense Lacan's theoretical figure of female *jouissance* as the ineffable and unknowable non-space *au-delà du phallus* derives from this earlier discourse.[11] The figure of the murderous erotomaniac is thus a point of intersection between the literary and the pseudo-scientific. It shows to what extent the pseudo-scientific is imbricated in the literary. Indeed, the celebration of the ecstatic female become transgressive may be as much informed by the Gothic and late-Gothic tradition of 'la donna delinquenta' as by psychiatric and psychoanalytic case studies. The work of writers like Krafft-Ebing and Lombroso,[12] classifying sexual aberrations and criminal types according to dubious scientific codes, bears out to what extent pseudo-science takes its material from the literature of terror. From one reading Bataille's Simone and Marcelle are positioned thus at the

[7] Cf. Georges Didi-Hubermann, *Invention de l'hysterie. Charcot et l'iconographie photographique de la Salpetrière* (Paris: Macula, 1882).

[8] The Papin sisters murder would be the subject of a study by Lacan in the review founded by Bataille, *Minotaure*—'Motifs du crime paranoïaque', *Minotaure*, 3/4 (1933)—as well as providing the basis for Genet's play *Les bonnes*, of 1947.

[9] Jacques Lacan, *De la psychose paranoïaque dans ses rapports avec la personnalité* (Paris: Seuil, 1975, orig. Le François, 1932.)

[10] Jacques Lacan, *Encore* (Paris: Seuil, 1975), 70.

[11] Ibid., 69.

[12] Providing us with a trajectory which leads from the guillotine through pre-analytic pesudo-science to contemporary French psychoanalysis, in a text entitled *The Skull of Charlotte Corday* (London: Secker and Warburg, 1995), Leslie Dick traces the fortunes of the infamous heroine's guillotined cranium to its position as *topos* for an anthropometric study by Cesare Lombroso, titled *La donna delinquenta* and translated as *The Female Offender*. The skull passed into the ownership of Prince Bonaparte, father of Marie Bonaparte, one of the founders of the French psychoanalytic establishment and 'expert' on female frigidity.

intersection of a plurality of discourses—the Gothic figure of the cruel woman, such as Clara in Mirbeau's *Le jardin de supplices*,[13] the pseudo-scientific notion of the transgressive erotomaniac and the murderous heroine of Surrealist and later psychoanalytic discourse. It would be wrong, however, to limit our consideration of Bataille's heroines to this type. Breton's idealism leads him to posit the existence of a sur-reality existing above and beyond 'the world', and Lacan's discourse also suggests an ambiguous formalization of a space 'beyond' as substance of the ineffable. Bataille's strategy, however, resists this tendency to formalize the space 'on the other side' of transgression. Transgression is rather an operation or a strategy of the cut, the interruption, which destabilizes the fixity and structural coherence of form. There is no space 'beyond' the Symbolic, but a hole which, through centripetal force, makes signifiers bleed into one another and draws them towards its mouth.

The mouth is an eye. What may establish the proximity between the Surrealist and psychoanalytic iconography of ecstasy is the image of the upturned eyes, the whites of the eyes as the pupil turns upwards to the sky, to the sun. In the *récit* of *Histoire de l'œil* the image appears on the eyes of the female penitent who emerges from the confessional: 'le visage pale et exstasié: ainsi la tête en arrière et les yeux blancs révulsés . . .'.[14] The image appears and reappears, in Dalí's collage, in Buñuel and Dalí's *Un chien andalou* (this time it is a male subject in erotic beatitude as he caresses the breasts of the object of his desire), in Breton's *Nadja* (here again it is Paul Eluard in whose face the disturbing whiteness of the eyes strikes us),[15] in the eyes of Bernini's Saint Teresa adorning the cover of Lacan's *Encore*. But the white of the eye is also what establishes its leakage into the egg, the bull's testicle, the saucer of milk. This eye does not see; it is the eye becoming

[13] Octave Mirbeau, *Le jardin des supplices* (Paris: Gallimard, Coll. Folio, 1988).
[14] *O.C.* I, 60.
[15] Breton, *Nadja*, 28.

pineal. It is not the sign of an interior vision or a mystical illumination, but a cut, a hole in the head which radically destabilizes the body as a form of knowledge. The two images of Marcelle's dead eyes, and the priest's eye, which *is* Marcelle's eye, in Simone's sex are eyes disconnected from their capacity to operate as receptors for perception which can then be translated into knowledge. They are images of the eye as hole in which the hole does not offer access to an interior realm or a space beyond. The eye as hole is also no longer able to operate as part of any dialectic of looks or strategy of recognition. It is, however, an eye which looks, in the sense that Lacan writes of the blindspot, the *tâche aveugle* which sees me from the point which I am not looking at, and in the sense that Olympia holds us in her gaze. This structure of anamorphosis, where death looks at us, irredeemably ruins the dialectic of vision. *Les yeux révulses* do not see, do not give access to the divine, but they look at us according to a force of fascination which attracts the gaze to that point where the structure of the gaze implodes. The white eye, the eye as mouth, is also that hole which leads to the destructuring of the visible contours of objects, and words. It is that mouth which in its voracity pulls the eye out towards the egg, the testicle, milk, sperm, which pulls *œil* to *œuf*, *couille* . . .

A different psychoanalysis informs the image of the whites of the eyes, which appears twice in the context of Bataille's own psychoanalysis from 1926 to 1927 by Dr Adrien Borel. One such image is a photograph given by Borel to Bataille in 1926, showing the execution by mutilation, while still conscious, of a Chinese man — 'le supplice des cents morceaux', whose eyes turn upwards at moments of intense pain.[16] The second is the image of his own father which the writer of 'Coïncidences' describes:

> Mais le plus étrange était sa manière de regarder en pissant. Comme il ne voyait rien sa prunelle se dirigeait très souvent en haut dans le vide, sous la paupière, et cela arrivait en partic-

---

[16] These photographs are reproduced in Michel Surya, *Georges Bataille, la mort à l'œuvre.*

ulier dans les moments où il pissait. Il avait d'ailleurs de très grands yeux toujours très ouverts dans un visage taillé en bec d'aigle et ces grands yeux étaient donc presque entièrement blancs dans les moments où il pissait, avec une expression tout à fait abrutissante d'abandon et d'égarement dans un monde que lui seul pouvait voir et qui lui donnait un vague rire sardonique et absent...[17]

Whether Bataille's fascination for the photograph of the Chinese victim, which he later admits haunts his writing, is a 'symptom' of his anxiety in relation to the image of his blind father's eyes, or whether the image of his father's eyes is constructed, fictionalized as traumatic origin 'after the event' by the vision of the photograph, it is evident that the figure of the 'whites of the eyes' which determines the whole *récit* (producing the association with eggs, for example, but also with milk, and with the bull's testicles) functions according to the operations of the Freudian dreamwork, 'travail du rêve'. One might postulate that the *displacement* of objects along the chain (saucer of milk, egg, bull's testicles . . .) is determined by a turn *away* from the traumatic image of the father's blind eye, or that the image of the eye of 'Marcelle' in Simone's sex, 'en pleurant des larmes d'urine', *condenses* different elements associated with these two images—blindness, the eye which does not see, urination, *jouissance*, mutilation. The images of *Histoire de l'œil* would thus function *symptomatically* in relation to the traumatic vision of these two images.

But do we want to locate such images as the 'origin' of the text? To do so would be to fix its structural play in relation to what Derrida refers to as 'une immobilité fondatrice et une certitude rassurante',[18] however traumatic. I have also proposed a number of other figures as in some sense generative of the play of the *récit*: the eye as figure, the cut, the apparatus... And Bataille himself, in his re-reading of his own texts, proposes a number of different 'traumatic origins' of differing kinds: the image of the 'jésuve', the 'saillie anale' of

[17] *O.C.* I, 76.
[18] Derrida, 'La structure, le signe et le jeu', 410.

a monkey in London Zoo, the pineal eye, and so on. Which trauma should we choose? The necessity of choosing becomes redundant if we consider all of these 'origins' not according to their substance, to *what* they represent, but as formal or structural operations. The figure of the 'whites of the eyes' is thus the symptom of an operation through which vision (of *this* world) is denied, an operation of the removal of sight. And it is the operation which is foremost; not blindness as such, but the limit between seeing and not seeing, the transgressive passage from one to the other, the *cut* of this limit.

22. / *corrida!*

Histoire de l'œil organizes its narrative in relation to two distinct topographies. The first is the oneiric and unnamed 'village of X', consisting of cliffside, country road, 'maison de santé/chateau hanté'. Part Gothic nightmare, part domestic drama, this oneiric topography is precisely unlocated: the different scenes do not communicate with each other according to an obviously identifiable route, but are as if suspended in an indistinct space, bleeding into each other according to the oneiric mechanisms of displacement. However, the stagings of the second part of the *récit* are developed according to a precise location, that of the arena of Madrid and the 'church of Don Juan' in Seville, again not linked in a traceable way, but situated geographically. These two locations allow a relation to two important intertextual and historical seams which *Histoire de l'œil* will deploy—the culture of *tauromachie* and the mythology of Don Juan. Both are readable as part of a generalized exploitation of the *topos* of Spain as a space displaced from France. Spain is the country of action, France of reflection; Spain is the space where transgressive acts forbidden or unrepresentable on the French stage, on the stage of France, are realized. Molière, Mozart, Da Ponte, Mérimée, Beaumarchais, Malraux, Montherlant, for example, exploit this trope. Spain as the *arena* for the staging of death. In Bataille's œuvre the 'church of Don Juan' in Seville is a negative image of *Notre Dame de Reims*, the subject of Bataille's first text, the incarnation of the French nation versus its disincarnation, the profanation of the host versus the communion of the youth of France. Other Spanish locations feature in Bataille's imaginary: the volcano of Monserrat, Tossa del Mar, where Masson and Bataille toast the birth of *Acéphale* to the accompaniment of the overture to *Don Giovanni*, the Barcelona of *Le bleu du ciel*, 'Espagne libre', the subject of a prototype for the review *Critique*; Bataille's textual space organizes itself around Spain as the stage of death.

The arena is thus the space of the staging of death and sacrifice. *Histoire de l'œil*'s deployment of the *topos* of bullfighting is recognizable as part of a tradition that includes Montherlant, Hemingway (whom Bataille mentions in 'Réminiscences'),[1] Leiris, Picasso, and Masson; and with the last three the corrida becomes eroticized, the spectacle of sacrifice is represented as a symbolic version of the sexual act. Michel Leiris's later text *Miroir de la tauromachie*[2] is perhaps the clearest exposition of the eroticization of the *corrida* and its *metamorphosis* of the sexual act. Like *Histoire de l'œil*, it is illustrated by Masson. But Leiris's text is also part of an intertextual network which *Histoire de l'œil* inaugurates, and which features a play of references and dedications between Leiris and Bataille. Bataille's pedagogical study of eroticism of 1957, *L'érotisme*,[3] is dedicated to Leiris's *Miroir de la tauromachie* and Leiris's *L'âge d'homme*, which at one point he considered rewriting 'in the style of' *Histoire de l'œil*,

---

[1] *O.C.* I, 606: 'M'avisant d'un rapport de la scène à ma vie réelle je l'associa au récit d'une corrida célèbre, à laquelle effectivement j'assistai—la date et les noms sont exacts, Hemingway dans les livres y fait à plusieurs reprises allusion . . .'

[2] Michel Leiris, *Miroir de la tauromachie* (Paris: G.L.M., 1938).

[3] Bataille, *L'érotisme*, 5, 11.

announces its debt to Georges Bataille, 'qui est à l'origine de ce livre'.[4]

It will be illuminating to refer briefly to Leiris's *Miroir de la tauromachie*, and then to read the *topos* of bullfighting in *Histoire de l'œil* in relation to it. Bemoaning the absence of the sacred in the modern world, Leiris proposes that erotic activity be considered as tangential to the art, or rite, of bullfighting, as its *mirror*. Bullfighting, in revolving around the climactic moment of the 'passe' as the torero magically deflects the bull's horns away from their target, plays out a dynamic between death, or sovereignty, in Bataille's schema, and illusion, 'la feinte'. And this dynamic hinges on the *muleta*, the red cape which the torero uses. The position of this piece of red cloth in the *corrida* is precisely at the paroxysmic site of the *passe*, where the bull's horn is deflected away from the body of the *torero*. It is the object of simulation or illusion, under or behind which death is duped. In Leiris's *miroir*, the *muleta* is subtly associated with the female sex: it is a 'guenille éclatante', and André Masson's 'Tauromachies' drawings, which appear in the 1964 re-edition of the book, literalize this parallel between *corrida* and sex. In the *corrida* the *muleta*, then, is both the veil which simultaneously invites and hides death, and a figure for female genitalia. In *Histoire de l'œil* Bataille exploits the resonances of this symbolic reflection which links death to the sexual act, describing the corrida, like Leiris, as a metamorphosis:

> Il faut dire de plus que si, sans long arrêt et sans fin, le taureau passe et repasse brutalement dans la cape du matador, à un doigt de la ligne érigée du corps, n'importe qui éprouve la sensation de projection totale et répétée, particulière au jeu du coït. [. . .]
> Une étoffe d'un rouge vif et une épée brillante—en face d'un taureau qui agonise et dont le pelage est fumant à cause de la sueur et du sang—achèvent d'accomplir la métamorphose et de dégager le caractère le plus captivant du jeu.[5]

[4] Leiris, *L'âge d'homme*, 7.
[5] *O.C.* I, 50, 52.

The moments of the *passe* are described in detail, emphasizing the red *muleta* as the centre of the action:

> [. . .] Granero ayant pris le taureau, le combat commença avec brio et se poursuivait dans un délire d'acclamations. Le jeune homme faisait tourner autour de lui la bête furieuse dans une cape rose; chaque fois son corps était élevé par une sorte de jet en spirale et il évitait de très peu un choc formidable. A la fin, la mort du monstre solaire s'accomplit avec netteté, la bête aveuglée par le morceau de drap rouge, l'épée plongée profondément dans le corps déjà ensanglantée; une ovation incroyable eut lieu pendant que le taureau avec des incertitudes d'ivrogne s'agenouillait et se laissait tomber les jambes en l'air en expirant.[6]

Subsequently, in *Histoire de l'œil*, the metamorphosis will be acted out by the narrator and Simone, who leave the arena for a sordid and stinking back room, where they copulate. The strict symbolic translation between the bullfight and the sexual act seems at this point to be exceeded by the potential associations which the text proposes. The narrator thrusts 'dans [la] chair couleur de sang et baveuse d'abord mes doigts puis le membre viril lui même'.[7] If we attempt to read the correspondences between the bullfight and the sexual act we in effect encounter a problem. According to a strictly linear interpretation, the penis would correspond to the sword of the torero plunging into the flesh of the bull, but the vagina would also correspond to the *muleta*, in which case the bull's horn also takes on a phallic function, and the bull would also appear male if we consider it the object of seduction with the red *muleta* a substitute for the vagina. The torero is also endowed with both feminine and masculine characteristics, as connoted by the following description:

> Le costume du matador est expressif, parce qu'il sauvegarde la ligne droite toujours érigée très raide et comme un jaillissement

[6] Ibid., 53.
[7] Ibid., 54.

> chaque fois que le taureau bondit à coté du corps, et parce qu'il
> adhère étroitement au cul.[8]

With Granero's enucleation the gender opposition is scrambled
further: the bull's horn, phallic and aggressive, penetrates the
eye socket, which assumes the female, passive position:

> sur [la] balustrade les cornes frappèrent trois coups à toute
> volée, au troisième coup une corne défonça l'œil droit et toute la
> tête.[9]

But at the same time, the eye is described as 'thrusting' out
of its socket, becoming phallic:

> l'œil humain [. . .] avait jailli hors du visage de Granero avec la
> même force qu'un paquet d'entrailles jaillit hors du ventre.[10]

The series of oppositions and the way I have characterized
them above might result in the following semiotic rectangle:

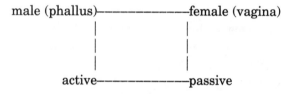

The opposition between genders is founded, in the above
account, on that between the phallus and the vagina, or that
which has phallic characteristics and that which has the
character of a hole; but this opposition is associated with
another, between active and passive. This tells us that the
whole structure is deployed around an *action* which constructs
an opposition between an actant and the object of an action.
The stability of the interpretation of the bullfight in relation
to gender roles conceived according to the active/passive opposition is ruined by a constant switching between the active
and the passive role. At one moment the bull is the passive
object of the thrust of a sword, the passive (male) object of

[8] Ibid., 52.
[9] Ibid., 56.
[10] Ibid., 57.

/ corrida!

seduction, at another he (?) is the active subject of an aggressive thrust.

In effect, Bataille's use of verbs plays with this opposition; for the effect of an action upon an object—the thrust of the bull's horn into Granero's eye, for example—an active verb ('jaillir') is used. For action, here, is not justified by intention: it functions intransitively, so to speak. *Histoire de l'œil* exploits the virtualities of the intransitive: 'ça jaillit'. This has the result of turning the gender opposition into a purely structural, permutational one, structured and destructured either side of the movement of rupture.[11]

The intransitive actions which generate the play of Bataille's description of the bullfight are all characterized by a sense of rupture, of movement in or movement out. They take place around a tear, a cut, or an opening. The narrator thus lists the elements of the bullfight which fascinate Simone in a way which draws attention not to the total spectacle of the bullfight as such, but to key moments which exploit this movement of rupture or transgression.

> Il y avait trois choses dans les corridas qui captivaient [Simone]: la première quand *le taureau débouche en bolide du toril ainsi qu'un gros rat*; la seconde *quand ses cornes se plongent jusqu'au crâne dans le flanc d'une jument*; la troisième quand cette absurde jument efflanquée galope à travers l'arène en ruant à contretemps et *en lâchant entre ses cuisses un gros et ignoble paquet d'entrailles aux affreuses couleurs pâles, blanc, rose et gris nacrés*.[12]

Déboucher, (se) plonger, lâcher: movements which produce and are at the same time constructed around an opening. It is this sense of rupture, the cut, which provides the association with the sexual act. The opening, the cut, of orgasm is a tear in the closure of the individual, whether male or female:

[11] Rosalind Krauss's description of Giacometti's *Suspended Ball* describes a similar effect, and she uses the enucleation episode from *Histoire de l'œil* as a parallel. See p. 21.
[12] *O.C.* I, 50 (my italics).

> L'orgasme du taureau n'est pas plus fort que celui qui nous arracha les reins en nous entre-déchira sans que mon gros membre eût reculé d'un seul cran hors ce cette vulve emplie jusqu'au fond et gorgée par le foutre.[13]

In Leiris's account, the *muleta* functions, in the movement of the *passe*, to disguise the lack of rupture. It is a *feinte*, the transgression is only feigned, deflected at the last moment. It is the proximity of this feigned rupture to the real one which provides the excitement of the bullfight. Bataille, less concerned than Leiris with the importance and the ubiquity of artifice, wants a *real* transgression. The bullfight, as such, is essentially boring:

> Sir Edmond et moi commencions à être ennuyés d'attirer l'attention de nos voisins juste à un moment où la course languissait.[14]

and:

> Il est juste de dire que, de plus, la course était devenue ennuyeuse, des taureaux peu combatifs se trouvant en face de matadors qui ne savaient pas comment les prendre . . .[15]

The bullfight needs to be supplemented by the non-feigned rupture, by a real sexual act, a real death: the penetration of Simone by the narrator, her insertion of the bull's testicle into her sex, Granero's death by enucleation. Bullfighting does not interest Bataille as such, then. It is rather the possibilities it offers of a failure of the feint that fascinate him.

Moreover, a further set of concerns inform the presence of the bullfight in *Histoire de l'œil*, concerns in which the symbolic and mythical functions of the bull and of the sun are exploited.

There is a marked insistence in the description of the bullfight scene on the character of the light. It is a particularly *Spanish* light:

> Il faut tenir compte aussi du ciel torride particulière à l'Es-

---

[13] Ibid., 54.
[14] Ibid.
[15] Ibid., 55.

pagne, qui n'est pas du tout coloré et dur comme on l'imagine; il n'est que parfaitement solaire avec une luminosité éclatante mais molle, chaude et trouble, parfois même irréelle à force de suggérer la liberté des sens par l'intensité de la lumière lieé à celle de la chaleur.[16]

I have already noted the exploitation of the association of light and liquidity, present here. The *particularity* of this light, which, as the narrator tells us, is almost all he has conserved of the memory of the events, suggests that the light of the sun is also an essential element of the spectacle of the bullfight, part of its drama. The oppressive quality of the sun is proposed several times. The narrator, Sir Edmond and Simone have seats 'en plein soleil'.[17] They are 'pris dans une sorte d'immense buée de lumière et de chaleur moite qui désséchait la gorge et oppressait',[18] and absorbed by 'le rayonnement solaire'.[19] The light of the sun, moreover, is described as 'aveuglant',[20] and this functions as a link to the next scene, in the chapter 'Sous le soleil de Séville'. The Sevillian light and heat are described as 'encore plus déliquescentes qu'à Madrid',[21] and this excess parallels the excessive step which the trio take in enucleating and blinding (*aveugler*) the priest in the 'church of Don Juan'.

The sun, therefore, has a particular function in the narrative. The 'liquéfaction urinaire du ciel'[22] at the moment of Granero's blinding would imply that it has a quite specific and active role in the structural play of associations which the text enacts. At this point it would be desirable to explore the place and function of the sun throughout the work of Bataille, to explore what position it had in the economy of signifiers which his writing exploits and constructs. The task would be an immense one, however, and we are limited to mentioning a

[16] Ibid., 52.
[17] Ibid., 54.
[18] Ibid., 55.
[19] Ibid., 55.
[20] Ibid., 56.
[21] Ibid., 57.
[22] Ibid., 57.

few factors of the *solarity* of Bataille's imaginary. *L'anus solaire*, written before *Histoire de l'œil*, would obviously be a key text. It includes the following fragmentary *enoncé*:

> Les yeux humains ne supportent ni le soleil, ni le coït, ni le cadavre, ni l'obscurité, mais avec des réactions différentes.[23]

The entry for 'Soleil pourri' in *Documents* offers the following phrase:

> Le soleil, humainement parlant (c'est à dire en tant qu'il se confond à la notion de midi) est la conception la plus *élévée*. C'est aussi la chose la plus abstraite, puisqu'il est impossible de la regarder fixement à cette heure-là.
> [. . .]
> Mythologiquement, le soleil regardé s'identifie avec un homme qui égorge un taureau (Mithra) [. . .] celui qui regarde avec le taureau égorgé . . . le taureau lui-même est aussi pour sa part une image du sang, mais seulement égorgé.[24]

In Mithraic cults, the sacrifice of the bull symbolizes the gift without return or exchange of the sun's light. The blood of the bull would stand for the light of the sun, the bull for the sun itself, but only at the moment of its death.[25] The bullfight in

[23] Ibid., 85.

[24] Ibid., 231.

[25] Another series of symbolic resonances may have been exploited here, associated with the figure of the minotaur. I will give only a few indications: the minotaur is a monster with the body of a man and the head of a bull, a headless, acephalic man. (This figure would inspire the journal and 'secret society' which Bataille formed in 1937, *Acéphale*. *Minotaure* would also be the title of a journal in whose foundation Bataille was involved. It would be 'taken over' by Breton and the Surrealists.) The minotaur is the progeny of Pasiphaé, the wife of King Minos, whose curse was to desire to copulate with a bull. The minotaur is hidden in a labyrinth, constructed by Daedalus, which Theseus finds his way through with the aid of Ariadne's thread. But as a result of Theseus's escape Daedalus and Icarus, his son, are imprisoned in the labyrinth on the order of Minos. Escaping with the aid of a pair of wings made by Daedalus, Icarus flies too close to the sun and both Daedalus and Icarus fall into the sea. The labyrinth, Ariadne's thread, Icarian flight, the acephalic man-bull are figures which re-occur throughout Bataille's early work, the common element being, in my reading, the sun as a principle of *dépense*. A reading of Racine's *Phèdre* tells us that the whole story is caused by a curse put on the daughters of the sun by Athena, who wished revenge for the sun having revealed her nakedness.

Histoire de l'œil thus serves another function, distinct from the 'sphère métaphorique' detected by Barthes (in which the sun does not feature) which is to represent the gift of the sun's light, the condition of vision. The death of the bull killed by Granero is in effect rendered by 'la mort du monstre *solaire*'.[26] The bull 'is' the sun, according to this representation of sacrifice, and as such a necessary factor of the bullfight-sacrifice is an extreme 'deliquescence'[27] of the sun. With the death of Granero a reverse image of the sacrifice is given. The torero's sword, instrument of its death, becomes the bull's horn, the eye the figure of the sun which is mutilated, ruptured. This figural representation of the sacrifice of the sun coincides with the liquefaction of the sky—light figured as liquid which bursts out of the sun and ruptures its circularity.

The sun, in Bataille's universe, gives its light *without reserve or return*, in an originary *dépense*. It is at the same time the most elevated conception and that which, as the cause of light and the condition of vision, cannot be looked at. Its sight is as scandalous and as insupportable as darkness or the sexual act. To look at the sun is to become blind, and since the sun is the condition of seeing, this blindness would also be linked to the principle of visibility. Blindness is thus the result of a transgressive look at the very possibility of the look, the result of a tautological impossibility of seeing that which makes seeing possible. According to this reading, the 'scenes' of *Histoire de l'œil* would be structured according to a chain of *equivalents* for the figure of a fixed stare at the sun— scandalous visions of that which human eyes cannot bear. But as a representation, this *equivalent* would necessarily be a trope, a rhetorical figure of displacement, a turning away. The *equivalent* is thus never the sign of equivalence, of the Same;

[26] *O.C.* I, 53.

[27] The bullfight scene is described as 'une simple vision de la déliquescence solaire' (53), and the state of the three figures as a shared 'déliquescence morose' (55). Deliquescence: from the Latin *delict*—to violate. The figure which generates this scene is thus the *violation* of the circular enclosure of the sun by the excess of its light.

the copula is figural, undermined by the movement of the turn. The turn, the trope, would also be a movement *away* from the impossibility of confronting the 'origin', an origin which is original only in its difference, only in the movement of the turn. The 'origin' in this sense would also be the condition of possibility. The story of the eye would be a narrative generated by the desire to find the equivalent of staring at the sun— tautological figure of vision confronting what makes it possible—and by the necessity of only ever arriving at displaced figures of that look. The 'origin' of the text, the hidden figure at its heart would thus be a figure of blindness, a tautological, fixed reflection in which no turn, no trope were possible.

Solar imagery does not in fact become an explicit concern in the text until the bullfight scene, until Spain. However, previously, the implosion of the 'œuf a demi gobé' occurs in the rays of 'le soleil oblique de six heures',[28] and Simone's associations with the word '*uriner*' had been: '*buriner*, les yeux, avec un rasoir, quelque chose de rouge, le soleil'.[29] Another game with eggs in, which the narrator and Simone imagine, involves the former shooting at eggs in the air 'au soleil'.[30] Eggs, close to the sun because of their shape and the colour of their yolk, function thus as displaced equivalents for the sun. The association of the sun with '*uriner*' looks forward to the 'liquéfaction urinaire du ciel', and combines with the association of the 'vase de nuit antique en terre poreuse abandonné un jour de pluie d'automne sur le toit de zinc d'une buanderie provinciale',[31] to produce, at the end of the text, a description of Andalucia as 'immense vase de nuit inondé de lumière solaire où je violais chaque jour [Simone] surtout vers midi en plein soleil'.[32] The urine 'usually' contained in the 'vase de nuit' is replaced first by rain, then by the yellow light of the Spanish sun which inundates, especially at noon. Evidently, the series

[28] *O.C.* I, 37.
[29] Ibid., 38.
[30] Ibid.
[31] Ibid., 32.
[32] Ibid., 69.

of associations at work here is inexhaustible, but one might point to the Freudian notion of *overdetermination* to account for the extremely oneiric and complex image of this 'vase de nuit'.

While the sun appears explicitly only in the Spanish chapters, it is however 'invisibly' present earlier. One of the earlier chapters is titled 'Une tache de soleil';[33] it describes the visit of Simone and the narrator to the asylum where Marcelle is imprisoned, and their vision of her window as a 'trou rectangulaire perçant la nuit opaque'[34] and 'le trou éclairé de la fenêtre vide'.[35] The image of the lit window is another variant of the blinding sun, as is also the moon ('le disque lunaire').[36] The 'stain on the sun', or 'sunspot' ('tache de soleil') may be produced by the reversal of the image of the sun. The darkness is the sun (like the sun it is invisible—Bataille would later cite Nietzsche's 'La nuit est aussi un soleil')[37]—the lit window a stain or spot upon it. The image is reproduced, moreover, in the stain of urine (again anticipating the association of urine and sunlight) upon the sheet which Marcelle hangs out of her window.

A further particularity of the imagery of the solar: on two occasions the same image is repeated. At the bullfight, Simone and the narrator 'make love' in a stinking toilet in which 'des mouches tourbillonait dans un rayon de soleil';[38] and later, in the 'church of Don Juan' (in a chapter titled 'Les pattes de mouche'): 'une mouche [. . .] bourdonnait dans un rayon de soleil et revenait sans cesse se poser sur cette figure'.[39] The fly

---

[33] Bataille's chapter titles, such as 'Une tache de soleil' or 'L'œil de chat', have the apparent function of indicating the hidden, or not yet realized, or intertextual associations of the text, the latter anticipating the *Documents* entry on 'Œil', the former suggesting the later entrance of the sun as a primary figure and its association with urine.

[34] *O.C.* I, 31.

[35] Ibid., 32.

[36] Ibid., 580 (second version).

[37] Used as the epigraph to *L'expérience intérieure*.

[38] *O.C.* I, 54.

[39] Ibid., 67.

will then land on the eye of the corpse (of the priest): 'Voilà ce qui était arrivé de bizarre et complètement confondant: la mouche était venue cette fois se poser sur l'œil du mort et agitait ses longues pattes de cauchemar sur l'étrange globe. La jeune fille se prit la tête dans les mains et la secoua en frissonant, puis elle sembla se plonger dans un abîme de réflexions.'[40] The abyss of reflections into which the image of the fly's legs on the priest's eye plunges Simone would lead us first to the fly 'in' the sun, and then, through a kind of superimposition, to the 'sunspot' and the stain of urine on the white sheet. But one would also want to underline the force of the image of the eye 'touched' by the 'nightmarish' legs of the fly, a realization of the phrase 'toucher à l'œil', where the monstrous 'acts out' its affront to the visible. The 'abîme de réflexions', in this case the itinerary of the fly itself as 'cauchemar', would be intertextual, leading us through Bataille's work, from the mention in the 'Figure humaine' *Documents* article of the 'mouche sur le nez de l'orateur' which parodies the stable ego,[41] to the (uncommented-on) image accompanying the article 'L'esprit moderne et le jeu des transpositions': Boiffard's photograph of 'Mouche et papiers collés'.[42] In the later novel *Le bleu du ciel*, the image of the stain on the sun becomes part of another complex play of superimposition and displacement: the image of an aeroplane against the blue sky of noon, carrying the narrator's lover Dirty into Barcelona, is superimposable on an image of a fly drowning in a saucer of milk produced earlier in the text.[43] How might a commentary account for such a play of images which threatens to overflow the bounds of 'commentary', of 'imagery'? We might associate the fly, or the stain, with the spider or spit, that is, the *informe*, that which knowledge cannot accommodate and

[40] Ibid.
[41] Ibid., 183.
[42] Reproduced in *O.C.* I, Planche XXV.
[43] *O.C.* III, 415.

must annihilate. That a fly buzzes in the sun, or drowns in milk, that one spits in soup would suggest that the totality (the sun's sphere, white as the absence or plenitude of colour) is always going to be corrupted by a base materiality, that which cannot be integrated into any system—that which disrupts the system. The image of the stain might offer us an abstract version of this generative figure.[44]

Blindness ('aveuglement') is also a characteristic whose fortunes it is possible to trace in an explicit manner, while maintaining the notion of an implicit, structural function for blindness in relation to the text's rhetorical economy. The Phrygian bonnet worn by the narrator at the orgy and when he releases Marcelle from her prison is 'd'un rouge aveuglant'.[45] The state of mind in which Marcelle's death and her corpse leave the narrator and Simone has the character of rendering them blind, that is, as the text qualifies, 'situés très loin de ce que nous touchions'.[46] However, their blindness—a removal from 'this world'—is countered by the 'yeux chatrés' of those to whom the universe appears honest ('honnête').[47] The sun at the bullfight naturally leads to an 'aveuglement des yeux', and is described as a 'soleil aveuglant'.[48] One would also want to follow, in relation to the theme of blindness, the appearances of the fixed stare (e.g. 'nous ne nous regardons guère fixement qu'à des moments analogues',[49] 'ces régions déséspérantes que Simone [. . .] me laissait regarder comme en hypnose',[50] 'fixer

---

[44] The image of the fly in milk appears in Villon's *Testament*, a text with which Bataille may well have been familiar, in a poem titled 'Le Debat de Cuer et du Corps de Villon'. The head replies to the heart's statement 'Rien ne coignois' ('You know nothing'): 'Si fais, mouche en let. L'ung est blanc, l'autre est noir./ C'est la distance.' Villon's poem invites us to read Bataille's image of 'noir sur blanc' as suggesting a *base* form of human knowledge, a fundamental heterogeneity.

[45] *O.C.* I, 43.
[46] Ibid., 47.
[47] Ibid., 45.
[48] Ibid., 55, 56.
[49] Ibid., 15.
[50] Ibid., 26.

sur les *œufs* des *yeux* grands ouverts')[51] as variants of the tautological reflection of sun and eye.

Bataille's text eroticizes and aestheticizes the event of Granero's death, integrates it into a chain of associations, a permutative play.[52] Nonetheless, this is an event which is marked punctually as having happened, at a certain time—7 May 1922—and at a precise location—the Madrid arena. The punctuality of this event, which is reproduced in photographs, has an affective power beyond any cultural, aesthetic exploitation of bullfighting, it is a point at which 'the real' punctures the text like a bull's horn. The date 7 May 1922—is that *punctum* of the real which tears a hole in the screen of fiction. It happened, like the torture of the victim of the 'supplice des cents morceaux'. The status of Granero's death as an event situated both historically and geographically has the effect of confusing any distinction between narrator, writer and author. We know, as readers, that the bullfight at which Granero was killed was a real event witnessed by Bataille while at the *Ecole des hautes études hispaniques* in Madrid, in 1922. This does not however lead us to assume that *all*, or even any other of the events described in *Histoire de l'œil* are ones which Bataille played a part in or experienced directly. On the other hand the status of this event means that we cannot read it solely on the level of the imaginary. It complicates the distinction Barthes makes between the novel, unfolding in relation to the contingencies of the real, and the poem as 'exploration exacte et complète d'éléments virtuels'.[53] Granero's death becomes integrated with the structural play of images in the text, but the inscription of the date and the name, the historical *punctuality* of the enucleation, maintain the uncomfortable intrusion of the real. In effect this reflects a further complication: at certain moments the associations and obsessions which struc-

---

[51] Ibid., 37.

[52] The *aestheticization* of the bullfight is suggested by the description of the horse's entrails as *'affreuses couleurs pâles, blanc, rose et gris nacrés*, suggesting the bullfight paintings of Goya or Manet. See p. 113 earlier.

[53] Barthes, 'La métaphore de l'œil', 239.

ture the text are ascribed to the conscious or unconscious interiority of the characters within the *récit*. At other moments the narrator adopts a 'theoretical' voice which leads to a confusion between 'his' theory of eroticism and that of 'Bataille'. And we have also seen how other concerns are active in generating the text which I have ascribed to the aesthetic imaginary of Bataille as a writer, or an individual with certain experiences, as in the case of Granero's death. 'Coïncidences' also doubles the *récit* with an explanation of the genesis of the text's images in relation to the life of the author, further problematizing any clear-cut distinctions between 'fiction' and 'real'. We would not be justified in any circumstances in reading the text as an autobiography, but there is an extent to which Bataille, as a writer, inscribes his own obsessions and his own imaginary concerns into the tissue of the fiction, and takes on the voice of his narrator. The *distance* of the text as 'fiction' is in doubt. As such the indeterminacy of the *subject* of the imaginary images or symptoms which structure the text is something to be retained, rather than resolved.

It is in this light that we might read the following passage, which describes the narrator's memory of the bullfight:

> En fait cette extrême irréalité de l'éclat solaire est tellement lié à tout ce qui eut lieu autour de moi pendant la corrida du 7 mai que les seuls objets que j'aie jamais conservés avec attention sont un éventail de papier rond, mi-jaune, mi-bleu, que Simone avait ce jour-là et une petite brochure illustré où se trouve un récit de toutes les circonstances et quelques photographies. Plus tard, au cours d'un embarquement, la petite valise qui contenait ces deux *souvenirs* tomba dans la mer d'où elle fut retirée par un Arabe à l'aide d'une longue perche. C'est pourquoi ils sont en très mauvais état, mais ils me sont nécessaires pour rattacher au sol terrestre, à un lieu géographique, à une date précise, ce que mon imagination me représente malgré moi comme une simple vision de la délinquescence solaire.[54]

The brochure, the suitcase, the Arab, the pole *attach* the narrator's imaginary vision of solar delinquency to the historic

[54] *O.C.* I, 53.

and geographic real, to the ground. They transform something announced as a product of the imagination into a memory (*souvenir*—underlined in the text). The text thus comments its own *attachment* to the real, to the memory of an event and thus to the experience of the writer. Granero's death is the point at which the imaginary *récit joins* with the memory of its author and the reality of his experience. Moreover, the text comments its own dynamic: the integration of Granero's death into its structural, imaginary play, suggests that the punctuality of this event is threatened, almost effaced, by the unreality of the solar ('l'extrême irréalité de l'éclat solaire'). The brochure, suitcase, Arab with pole are *not* readable as part of any associative chain of imagery—they fulfill no function except that of suggesting the punctuality of the event and the irreducibility of its memory.

23. / don juan

After the death of Granero, the itinerary of the trio of erotic adventurers continues south, from Madrid, to the city of Seville, ostensibly because it is known as 'une ville de plaisir'.[1] This pleasure, which is ascribed in the text to the 'deliquescence'[2] of heat and light, and the enervating scent of flowers in the streets, is also overdetermined by the identification of Seville as the site of a certain *literary* version of pleasure. Seville is the city of Don Juan and Figaro; the plays, operas and poems celebrating these figures concern seduction and the pleasures of the flesh. Bataille signals this intertextuality with the naming of Seville as the place of publication given for the second edition of the book. The deliberate emphasis serves the function of pointing to the location of

[1] *O.C.* I, 57.
[2] Ibid.

Seville as crucial in both geographic and literary terms. But it is the myth of Don Juan which *Histoire de l'œil* will specifically take up: the church in Seville which the trio arrive at is, according to Sir Edmond, 'l'église de Don Juan'. Sir Edmond explains:

> En riant de plus belle, il désigna sous nos pieds une grande plaque funéraire en cuivre. C'était la tombe du fondateur de l'église, que les guides disent avoir été Don Juan: repenti, il s'était fait enterrer sous le seuil pour que son cadavre fût foulé aux pieds par les fidèles à l'entrée et à la sortie de leur repaire.[3]

The mention of the founder of the church apparently having been Don Juan, according to 'the guides', and further down of 'deux célèbres tableaux du peintre Valdès Léal',[4] have made possible the identification of this church as the Hopital de la Caridad in Seville.[5] The church was founded in 1676 by Don Miguel de Manara who according to popular legend was the inspiration for Byron's Don Juan. Manara is said to have repented a dissolute life after a vision of his own death and joined the Brotherhood of Charity, whose task it was to bury the bodies of criminals and the destitute. The hospital would be devoted to the care of the terminally sick. To decorate it Manara commissioned eleven paintings by Murillo and two by Valdès Léal, the latter showing two images of death. One shows the decomposing corpse of a bishop with above it a divine hand holding a pair of scales. This is the painting to which Bataille refers, although the other painting, showing a figure of death as skeleton with scythe, includes the inscription 'In ictu oculi' ('in the blink of an eye'), a device which signals the temporality of the *cut* I have already discussed.[6]

---

[3] Ibid., 59.
[4] Ibid.
[5] Cf. Chakè Matossian, 'Le rat et l'œuf (Bataille, l'*Histoire de l'œil* et le clin d'œil à Valdès Léal)', in *La part de l'œil*, 10 (1994): 'Bataille et les arts plastiques'.
[6] Chakè Matossian reads Bataille's choice of the Valdès Léal paintings instead of the Murillos as a deliberate iconographic choice for the morbid over the harmonious, and suggests that *Histoire de l'œil* is partly a counter to Murillo.

Manara, the founder of this church, is evidently not the Don Juan of Tirso de Molina or Molière, or Mozart, but that of Byron, a Don Juan who repents. The incidence of the 'church of Don Juan' in the text is specifically not part of the literary mythology of the stone guest and the Commandatore, or the crack which opens in the earth into which Don Juan descends. This is the kind of intertextual association one would expect to find, given Bataille's later celebration of the figure. The myth of Don Juan will serve Bataille for a number of later works: the Commandatore makes an appearance in *Le bleu du ciel*; the birth of the review *Acéphale* is accompanied by the overture to Mozart's *Don Giovanni*, a record of which Masson plays to Bataille in Tossa, Spain. The promise of Don Juan as a *scandalous* literary figure fails here; the presence of Don Juan is determined by the contingency of the place, and of the (erroneous) association by the 'guides' of Miguel de Manara with the legendary seducer.[7]

I suggest that something else is going on here other than the purely accidental detail of the place of the Hopital de la Caridad and the Valdès Léal paintings, although one suspects that, like the death of Granero, this is a part of the text which is again corrupted by the contingency of the real, of Bataille's experience. The disappointment of the expectation of a legendary and scandalous Don Juan sets up a desire for some *other* determination, and one wants to look elsewhere for an explanation of the relevance of this site and the figure of Don Juan.

Description is not a mode which *Histoire de l'œil* adopts,

She reads the textual and iconographic associations of the two Valdès Léal paintings against Bataille's text to produce a commentary on the instantaneity of the moment of transformation which is death.

[7] The contingency of this association is emphasized by the following entry in a contemporary guide to Seville, the *Rough Guide to Andalucia* (London: Penguin, 1994): '. . . the Hopital de la Caridad, founded in 1676 by Don Miguel de Manara, who may well have been the inspiration for Byron's Don Juan . . .' (182).

for the most part. As such it is significant that the church (not named in the text) is afforded a rich descriptive account:

> Nous nous trouvâmes dans une salle où nous ne vîmes rien qui justifiât le rire de Simone; relativement fraîche, elle recevait la lumière à travers les rideaux de cretonne rouge. Le plafond était fait d'une charpente ouvragée, les murs blancs, mais ornés de statues et d'images; un autel et un dessus d'autel dorés occupaient le mur du fond jusqu'aux poutres de la charpente. Ce meuble de féerie, comme chargé des trésors de l'Inde, à force d'ornements, de volutes, de torsades, évoquait par ses ombres et l'éclat des ors les secrets parfumés du corps. A droite et à gauche de la porte, deux célèbres tableaux de Valdès Léal figuraient des cadavres en décomposition: dans l'orbite oculaire d'un évêque entrait un énorme rat.[8]

The description is, like the church, 'sensuel et somptueux'; with its combination of red and gold it evokes immediately the literature of the Romantic era—the excessive but dark atmosphere, the suggestion of Orientalism, and the touch of terror marked by the paintings. The location, in other words, signals its appurtenance to the Romantic tradition of the 'tale of terror', already marked with the earlier 'chateau hanté'. In a sense, one might say that the description of the church is typically 'Balzacian', firstly in that its function, outside the logic of the narrative *per se*, is to connote that 'effet de réel' which Barthes signals as a key element of the realist text, and secondly in that Balzac's 'realism' is invariably complicated by the tradition of the supernatural and that of the literature of terror.

The reader will be aware at this point of being 'set up' to encounter a text by Balzac as a privileged intertext, or perhaps a veritable pretext, for *Histoire de l'œil*. An early text by Balzac, *L'elixir de longue vie*, parodically and sensationally treats the Don Juan myth, and provides Bataille's text with an already sacrilegious and scandalous approach to the Don

[8] *O.C.* I, 59.

Juan myth and with the genre of the tale of terror to depart from.[9]

This is the story: the dissolute Don Juan's aged father is on the point of death. At the final moment, Don Belvidéro calls his son to his deathbed, seemingly untroubled by his imminent end. 'Dieu, c'est moi' cries the old man in response to his son's (feigned) religious imprecations.[10] He tells his son where to find a small phial of an elixir which, when spread on his corpse, will bring him back to life. His father dies and Don Juan is 'perdu dans un monde de pensées'.[11] He realizes the advantages he might gain from the elixir, and, after his father's death, instead of spreading the elixir over the whole of the corpse, drops a tiny amount on to his father's eye. 'Imbibe un œil' his inner voice tells him.[12] The eye comes to life, and its menacing gaze follows Don Juan around the room. Once aware of Don Juan's parricidal desire to crush the eye ('le crever'), it cries burning tears: 'Une grosse larme roula sur les joues creuses du cadavre et tomba sur la main de Belvidéro. "Elle est brûlante!" s'écria-t-il en s'asseyant'.[13] Finally, Don Juan crushes the eye into the skull of the corpse.

Thereafter Don Juan lives the life of a libertine, knowing that his death can be postponed. He converses with the Pope and corrupts him. On his deathbed he calls his son to his bedside, telling him that the elixir will cleanse his body of the impurities of sin. At the given moment, the son begins to spreads the potion over his father's body, but, when the upper half of the body comes to life, he is so shocked that he drops the phial, which smashes on the ground. The recovery of the upper half of Don Juan's body is hailed by the church as a

[9] Honoré de Balzac, *Contes choisis* (London, New York: J. M. Dent, Coll. 'Les classiques français'). Steinmetz points to the Balzac text as 'une autre "histoire de l'œil"' in his 'Bataille le mithriaque', 182.
[10] Balzac, *Contes choisis*, 13.
[11] Ibid., 15–16.
[12] Ibid., 19.
[13] Ibid., 21.

miracle, and Don Juan canonized as a saint. The ceremony takes place at the Church of 'San Lucar' in Seville. Don Juan's body, the head and shoulders alive but the rest of the body a lifeless corpse, is positioned above the altar. At the climactic point of the ceremony, the head cries 'O coglione!' ('Balls'), and shouts a series of imprecations ('comme un ruisseau de laves brûlantes par une irruption du Vésuve'),[14] falls off the decomposing body and on to the head of the priest, where it gnaws into the brains of the unfortunate cleric.

If the element of the supernatural, which Balzac's later novel *La peau de chagrin* will integrate into the workings of a realist narrative, is replaced in Bataille's text by the limitless *vertige* of the chain of images, Balzac's text offers many points of proximity not only with *Histoire de l'œil*, but with Bataille's œuvre as a whole. The location of 'l'église de Don Juan' is common to both, as is the figure of the mutilation of the eye. It is a textual mutilation, a severing of the eye from the unified body and its autonomous functioning as an object and signifier. If Don Juan's mutilation of the father's eye in *L'elixir de longue vie* enables his liberation from the paternal law, the mutilation and *migration* of the eye as signifier in *Histoire de l'œil* can also be understood as a transgression of the law which regulates and resists the play of the signifier. Balzac's text also provides a rich material of obscenity and profanation for Bataille: Don Juan's expletive 'O coglione!' mimics Sir Edmond's 'Bloody girl', and, in the other Sevillian literary example, Figaro's *Goddam*, in *Le mariage de Figaro*.[15] The Balzacian image of the 'Vésuve' is resonant with the Bataillean

---

[14] Ibid., 39.

[15] The English expletives, in both *Histoire de l'œil* and *Le mariage de Figaro*, function in a similar manner, to tie the text to the literary context of England and primarily to the literature of terror which begins in the eighteenth century with the works of Radcliffe, Walpole and Maturin, and continues in the nineteenth century, particularly in the literature associated with what Mario Praz calls *le vice anglais*—the literature of erotic cruelty. It is significant that, as Praz points out, the evolution of the genre sees a shift of focus from a male hero, often a priest, to a female heroine.

image of the volcano as the sexualized abyss of the earth.[16] The phrase 'Dieu, c'est moi' is also echoed by Madame Edwarda's blasphemous statement. 'je suis DIEU' as she shows her 'guenilles velus et roses' to the narrator of that text.[17] But Bataille inverts the direction of the narrative: *Histoire de l'œil* ends, after its trajectory of signifiers liberated from fixity, with the literal enucleation of the priest, while Balzac's begins with enucleation and moves towards liberation, this time in a moral sense.

The importance of *L'elixir de longue vie* for *Histoire de l'œil* is that it provides, once more, the limitless and vertiginous text of the image chain with a narrative code, a law of genre, to work against. Such that the transgressive operation of Bataille's text, in this instance, is to be measured against the narrative code of the tale of terror not only of the Balzac text, but of the strong literary tradition to which it belongs. Functioning in the same way as Lautréamont's *Maldoror*, *Histoire de l'œil* 'underwrites' the narrative code of the tale of terror by pitting against it the force of the labyrinthine image-chain. What *Histoire de l'œil* effectively does to the tale of terror is to sexualize and to pathologize it—the element of the supernatural is relocated in the perverse sexual drives of the characters, thus signaling the more or less evident unconscious determination of the Gothic tale of terror by a repressed sexuality. And rather than the pure *evil* of the heroes and heroines of the Gothic tradition, *Histoire de l'œil* implicitly posits the psychiatric and psychoanalytic code of 'normality' as that which is transgressed.

The final scene of the text, as I have already proposed, realizes an intersection of a number of different associations of images. Without ignoring the evident force of blasphemy in this scene, in which the communion is re-interpreted as a sexual sacrifice, I want to focus on the final image. The priest

---

[16] Cf. 'L'anus solaire', *O.C.* I, 85, 86. Bataille creates the pun 'Le Jésuve': 'je suis le *Jésuve*, immonde parodie du soleil torride et aveuglant.'
[17] *O.C.* III, 20–21.

has been killed by asphyxiation while penetrating Simone, and Sir Edmond has cut out his eye. The narrator and Simone make love with the eye between them, before the former sits back to gaze upon the following scene:

> en écartant les cuisses de Simone qui s'était couché sur le côté, je me trouvai en face de ce que, je me le figure ainsi, j'attendais depuis toujours de la même façon qu'une guillotine attend un cou à trancher. Il me semblait même que mes yeux me sortaient de la tête comme s'ils étaient érectiles à force d'horreur; je vis exactement, dans le vagin velu de Simone, l'œil bleu pâle de Marcelle qui me regardaient en pleurant des larmes d'urine.[18]

The narrator has the sense of always having been waiting for this sight. In effect, the structural play of the text had set this image up for him:—this haunting as if from the future draws our attention to a sense of conflict between the character—the narrator and his consciousness—and the structural determination of the text.[19] If Simone is 'lost in an abyss of reflections', or if the narrator has the sense of having been fated to see what he sees, it is because they are caught in a

---

[18] *O.C.* I, 69.

[19] It would be possible to read the *récit* in psychological terms, as the haunting of the two characters, Simone and the narrator, by the death of Marcelle, which they partially cause. They are led through this haunting to try to expiate their crime, resulting in the subsequent murder of the priest. I have not emphasized this reading, preferring to focus on the structural, intertextual and textual functioning of the *récit*. But see Chakè Matossian's article for such an account. Bataille gives an account of the eye as '*œil de la conscience*' in the 'œil' entry to *Documents*, reading an image by Grandville and a poem by Victor Hugo based upon it: 'L'œil occupe même un rang extrêmement élevé dans l'horreur étant entre autres *l'œil de la conscience*. On connaît suffisament le poème de Victor Hugo, l'œil obsédant et lugubre, œil vivant et affreusement rêvé par Grandville au cours d'un cauchemar qui précéda de peu sa mort' (*O.C.* I, 188). Grandville's image, published in the *Magasin pittoresque* in 1847, is titled 'Crime et expiation', and shows a murderer pursued by a gigantic eye, which is graphically repeated across the drawing, falling into the sea where it metamorphoses into a shoal of fish, still pursuing the criminal. *Histoire de l'œil* would thus be more pertinently read as a *haunting by the eye*, a scene which Balzac's *Elixir de longue vie* had already provided. It is Marcelle's eye which haunts the narrator and 'returns', suggesting that haunting takes place at the level of the structural organization of the text, rather than in terms of the interiority of its characters.

---

textual network which programmes what happens. The story of the eye is one which is 'inflicted' on the characters of its *récit*; Bataille thus stages a dynamic relation between the structural and permutative—textual—quality of language, 'les besognes des mots', and the nature of 'fiction', as a removed representation.

The sense of haunting to which I referred above and which I have discussed earlier is also evident with the phrase '[comme] une guillotine attend un cou à trancher'.[20] The narrator has always been waiting for this scene as a guillotine awaits a head to be severed. Or—the image which haunts the text as its future is an image of decapitation, an image which *decapitates*, which makes one lose one's head, in the sense both of losing control, and of losing that part of the body which governs through reason. The guillotine has also haunted the text through the figure of Marcelle, herself haunted by the 'curé de la guillotine', and through the intertextual seams of the Gothic tale of terror and the revolutionary Terror. In this final scene the 'curé de la guillotine' finally loses his head, or at least his eye. But this is also an image which decapitates because it desublimates, it removes the eye from its position in the head, or at the head, and puts it 'back' in its position as *the body's eye*. The eye become head or the head as eye is cut off. If the narrator's eyes 'sortaient de la tête comme s'ils étaient érectiles à force d'horreur', it is because, becoming phallic, they also are subject to this decapitation, a castration of the eye. The cut of the eye removes it as the head, as the phallus, as transcendental signifier or general equivalent which organizes equivalence on the plane of the visual. 'True' vision, finally *seeing* Marcelle's eye and being looked at in turn by it, is possible only on condition of this cut which removes the eye of the head and installs the eye of the body.

[20] *O.C.* I, 69.

24. / cure

Histoire de l'œil, we might remind ourselves, consists of two parts, the *récit* and 'Coïncidences' ('Réminiscences' in the second version), explicitly designated as a 'Deuxième partie'. This second part begins: 'Pendant que j'ai composé ce récit en partie imaginaire, j'ai été frappé par quelques coïncidences et comme elles me paraissent accuser indirectement le sens de ce que j'ai écrit, je tiens à les exposer.'[1] It is announced that the 'je' which produces this part is the writer, the 'compositeur' of the *récit*. While we are justified to a certain extent in simply identifying this subject with Bataille, the use of the pseudonym Lord Auch and the fact that this second part is designated as part of the *Histoire* make such an identification problematic. As part of the *Histoire* the 'story' of the composition of the *récit* and the writer's interpretation of it would have a fictional status. 'Coïncidences' is neither wholly autobiographical nor wholly fictional. It complicates the framing of the text and situates its subject in a strange position as if on the limit between 'fiction' and the real. The second part, which recalls certain 'indirect' unconscious determinations of the *récit* is, moreover, not wholly situated either 'before' or 'after' the *récit*, but 'pendant'. It does not describe *directly* the obsessions which consciously led the writer to write what he did, but certain memories which lead him to reinterpret his own text as having a profound link to his own unconscious, to his own more or less buried experience, and as having been influenced by this unconscious determination. The second part is thus situated as part of the process of writing of the text, involving a re-reading of the text and a re-writing of it according to this revelation of its unconscious determination.

Initially the text is given a *cathartic* function: 'J'ai commencé à écrire sans détermination précise, incité surtout par le désir d'oublier, au moins provisoirement, ce que je peux être

[1] *O.C.* I, 73.

ou faire personellement.'[2] Certain statements around the text by Bataille invite us moreover to read the writing of the text as a part of a *personal* process of *cure*. In response to his brother Martial's horrified reaction to the suggestion that the description of Bataille *père* (as mad, syphilitic and blind) was authentic, Georges would write:

> Mais je veux te dire ceci dès aujourd'hui, ce qui est arrivé il y a près de cinquante ans me fait encore trembler et je ne puis m'étonner si un jour je n'ai pas trouvé d'autre moyen de me sortir de là qu'en m'exprimant anonymement. J'ai été soigné (mon état étant grave) par un médecin qui m'a dit que le moyen que j'ai employé en dépit de tout était le meilleur que je pouvais trouver. Tu pourrais le voir: je suis sûr qu'il te le redirait.[3]

[2] Ibid.

[3] Georges Bataille, *Choix de lettres* (Paris: Gallimard, 1997), 569. Martial Bataille wrote to his brother in response to an interview with Georges by Madeleine Chapsal published in *L'Express*. Chapsal had written of Bataille: 'Il est d'ailleurs prêt à donner sur ses origines des renseignements qui ne sont pas sans importance. Son père, tabétique, était paralytique général et devint fou. Sa mère perdit complètement la raison' (cited in Bataille, *Choix de lettres*, 568). Bataille would respond to Martial, asserting that he had not spoken about this to Madeleine Chapsal, but that '[Ta lettre] m'a d'autant plus désesperé qu'il y a un fond de vrai dans les allegations de *L'Express*' (568). Moreover, Bataille would write: 'Je n'en ai parlé (écrit) qu'il y a très longtemps, d'une manière anonyme (sans signes et sans donner aucun nom).' He would refuse to publish a rebuttal of the accusations in *L'Express*. Denis Hollier, in *La prise de la Concorde*, points out that Bataille's response to his brother (in a letter which may not have been sent) never actually contradicts the inauthenticity of the account of 'Coïncidences', but emphasizes instead the importance of the account for the *cure* which Georges was undertaking. However, eighteen years earlier, in the 'Préface à l'*Histoire de l'œil*' in *Le Petit*, by 'Louis Trente', we can read the following: 'Je me suis branlé, nu, dans la nuit, devant le cadavre de ma mère (quelques personnes ont douté, lisant les "Coïncidences": n'avaient-elles pas le caractère fictif du récit? comme la "préface" les "Coïncidences" sont d'une exactitude littérale: bien des gens du village de R. en confirmeraient la substance . . .' (*O.C.* III, 60). There follows an account of how, in the face of the German advance in 1915, the writer of this preface, with his mother, had abandoned his father in the 'village of N.' and how the father, blind, mad and suffering, died alone and 'abandoned' in the occupied village. A number of factors qualify the assertion of this 'literal exactness': 1) It is 'authored' by Louis Trente, who assumes the identity of the author of 'Coïncidences', i.e. Lord Auch. The dual pseudonym for the one identity emphasizes the artifice of the name as a fiction created by the 'real' author, who is

Histoire de l'œil—*récit*—would thus be part of a cure, linked to the psychoanalytic cure Bataille undertook from 1925 to 1926 with Dr Adrien Borel on the recommendation of his physician, a Dr Dausse. The writer of this second part would thus be identified as Georges Bataille. Martial Bataille's reaction to the account of 'Bataille' would suggest otherwise, but what seems important, whatever 'authentic' status is accredited to the description of the father, and however strictly we interpret the 'je' of 'Coïncidences' with Georges Bataille, is that the *récit* is interpreted as part of a process of cure by its 'compositeur', and that 'Coïncidences' reads the *récit* in relation to the unconscious of its writer and the functioning of this unconscious.

Bataille's psychoanalysis with Dr Adrien Borel begins in 1925 and is terminated in 1926. The texts previous to this encounter are either determined by the desire to contain and enclose an affective charge, such as the chaste and noble *Notre Dame de Reims* and the scholarly articles for *Aréthuse*, or failed texts such as an abandoned novel 'in the style of Proust',

---

unnamed (immediately above this part of the text we read: 'I gave the author of *W.C.* the pseudonym of Troppman'). 'Louis Trente' is no more the 'I' who writes here than is Lord Auch or Troppman. The structural play of pseudonyms thus suggests the reality of the experiences told by the unnamed 'I'. 2) A further displacement of names: village of R (in *Le Petit*)—village of X (in 'Coincidences')—village of N (also in *Le Petit*). The 'literal exactness' of the events is problematized by being in relation to three different places, and the reader is left with the enigma of determining which name is 'real', which 'fictional', deciding perhaps, since all three are not names but letters, that the 'names have been changed' to guard against identification with the 'real' place. This would contradict the intention of the assertion, however. The 'literal' (i.e. of the letter) exactitude of the accounts is, nevertheless, contradicted by the displacement of the letter: Troppman–Auch–Trente, X–R–N 3) '[C]omme la "préface" les "Coïncidences" sont d'une exactitude littérale': both texts are liminary, inscribed on the border of a fiction, neither wholly distant and removed from experience, neither wholly immersed in it. 'Bataille''s gestures here are important in asserting the irreducible reality of the *experience*, and inscribing this assertion in a problematic frame and an unstable field of displacement. We can never be sure that we are 'in' 'fiction' or 'in' 'reality', since the operation of the text installs a third term which undoes both: *writing*.

and the destroyed *W.C.* The writing of texts like *Histoire de l'œil* and the *Documents* articles is post-1926, that is after the psychoanalytic encounter which ostensibly liberates the writing process that at the same time is part of it.

What kind of cure? The emergence of psychoanalysis in France, as its historians have indicated, is closely tied in to aesthetic contexts.[4] We cannot therefore take the step of rigorously separating the literary practice of the writer from the scientific or clinical practice of the analyst, and this much is borne out by the context of Bataille's analysis. The psychoanalysis Bataille encounters is one which is explicitly 'unorthodox', or non-clinical, at least. Adrien Borel, Bataille's analyst, was one of the founder members of the Société Psychanalytique de Paris, but within this, part of the group known as 'chauvinist', distinct from the 'orthodox' tendency of Marie Bonaparte. He was one of a group of analysts who frequented artistic circles, and one of those analysts who 'specialized' in the analysis of literary figures. He knew Queneau and Breton, and would become the analyst of Michel Leiris between 1929 and 1935 on the suggestion of Bataille. In 1925, Adrien Borel and his associate Gilbert Robin would found *L'évolution psychiatrique*, evidence of how their practice and the theory which underpins it is not a strict application of Freudian theory but is also informed by the earlier, non-psychoanalytic considerations of madness and of cure. In the words of a recent historian of psychoanalysis in France, Borel's practice was characterized by the use of 'une manière très libre des règles freudiennes [. . .] Il se moque volontiers des psychanalystes "double crème", c'est-à-dire de ceux qui croient tout arranger par le divan.'[5] In that Borel's therapeutic practice was based on 'une écoute de la souffrance',[6] the spoken performance of the patient assumes an almost cathartic quality; written per-

---

[4] Cf., for example, Elisabeth Roudinesco, *La Bataille de cent ans* (Paris: Ramsay, 1982).
[5] Ibid., Vol. I, 358.
[6] Ibid.

formance too, then, was also considered in this light. The analysis, if this name can still be applied, worked through the confrontation of an obsessional sexuality by the force of a quasi-cathartic expression.

In 1925 Borel and Robin published a text which elaborates their practice, *Les rêveurs éveillés*.[7] It appeared in the same series as Blanche Reverchon's translation of Freud's *Trois essais sur la sexualité* and *Le rêve et son interpretation*, as well as Breton's *Les pas perdus*. Its authors develop a definition of what they call 'schizomanie', which results from a loss of contact with reality, and an inability to distinguish between imagination and reality. They elaborate how certain representations censored by the conscious mind become charged with non-expended energy and become stagnant 'blocks' within the mind. These are described as 'prodigieux amas d'images enfermées dans notre mémoire'.[8] All that is needed is a calm and comfortable place, and a liberation of the mind from any moral stricture, for these images to come forth (*'jaillir'*). The authors write of an 'affectivité profonde'[9] in terms proximate to Bataille's notion of 'une région profonde de mon esprit', mentioned later in 'Coïncidences'.[10] Throughout this theoretical elaboration, the author's reference is not so much Freud as Bergson.

Freud, whom Bataille first read in 1923, first became known in France via the psychology of Janet and the philosophy of Bergson. Bataille's encounter with Freud, both in terms both of the prevailing psychoanalytic culture of the time and of the philosophy which informs his analyst Borel, takes place through the intermediary of Bergson's psychological philosophy. For Bergson memory holds an important place as the link between matter, which consists of images, and mind, which deals with representations. In Borel and Robin's formu-

[7] Adrien Borel and Gilbert Robin, *Les rêveurs éveillés* (Paris: Gallimard 1925).
[8] Ibid., 5.
[9] Ibid.
[10] *O.C.* I, 75.

lation, then, the process of cure enables the representation of images which exist at a deep level ('une région profonde', perhaps) of the psyche, and which, unrepresented, function as blockages, leading to symptoms on the affective level. In Bergsonian terms, unrepresented images—matter—remain as solid blocks in the memory ('prodigieux amas') until their representation allows them to enter consciousness—the mind.

The psychoanalytic context which is most immediately available to Bataille is thus not the orthodox Freudian one dominated by the Oedipal complex, but one which focuses on stagnant psychic blockages and the means of their representation. The place of the Freudian unconscious is in effect taken on by memory, according to a Bergsonian schema, and the place of repression taken up by the notion of blockage. It would be anachronistic, therefore, to read 'Coïncidences' explicitly in relation to orthodox Freudian psychoanalytic theory, to emphasize the terms 'unconscious', 'repression' (which Bataille does not use in an explicitly Freudian sense). However, it is possible to see some parallels between the Freudian theory of the 'dreamwork' and the reading of his own text that Bataille proposes, while also retaining the resonance of the Bergsonian terms of memory and blockage.

If we set out to trace the mechanism of this cure, the first statement we come to tells us that the *récit* was written 'sans détermination précise', but with the desire to forget 'what I am or do'. The obscenity of the text, then, is intended to erase or purge the obscenity which dominates the author's personality and his deeds. It is without 'precise determination'. The remainder of the text, however, will contradict this initial statement and trace a number of ways in which the images and associations of the *récit* are revealed *after the event* to have been precisely, but indirectly, determined by specific events in the author's life. One kind of reading, which would see the text as influenced in a general sense by the life of the author, as an expression of his intention, is denied, and another kind of reading which reads *precise* determinations is offered in its place. This strategy of reading is similar to that which Freud proposes for the dream, in *The Interpretation of Dreams*: not to read the dream *symbolically*, as a whole, but to read its fragments, to read the precise determinations of specific images as opposed to the totality of the dream.[11] With this proximity to the Freudian theory and practice of the dreamwork in mind (I will return to it later), 'Coïncidences' takes on the reading of incidents of continuity between the tissue of the writer's unconscious and the manifest text of the *récit*. They are equally the result of certain 'réminiscences', a work of memory or anamnesis which reactivates the author's childhood experience. This anamnesis follows an itinerary through a number of points, as follows:

1. Leafing through an American illustrated magazine one day the writer comes across two images, one of the village where his family lived and the other of the nearby ruins of a medieval castle. The coincidence or juxtaposition of these two images activates a memory of a childhood experience linked to these ruins. A first association is thus produced (which the writer does not mention, but which as readers we notice) between the writer's childhood experience and the text: the proximity of the two images of the village and the castle corresponds to the topography of the first part of the *récit*, around the 'village of X' and the nearby 'maison de santé', which in the imagination of the narrator is transformed into a 'chateau hanté'. The itinerary of rememoration or of the process of analysis which links manifest text to dormant memories begins with the juxtaposition of the two images. It is significant moreover that one of the images shows a ruined castle, lieu par excellence of the Gothic imagination or of the Sadean orgy, which tradition the text will exploit (the walls of the castle of the writer's childhood are described as 'parfaitement romantiques').[12] The principal memory which is reactivated by the two images, however, is revealed subsequently: on a

---

[11] Cf. Sigmund Freud, *The Interpretation of Dreams* (London: Penguin, 1976), 178.

[12] *O.C.* I, 73–4.

midnight excursion to the ruined castle accompanied by his mother and some 'chaste' young girls, the narrator is surprised and terrified by the sudden appearance of a phantom, 'un fantôme blanc et extrêmement lumineux'.[13] Although the writer 'knows' the phantom to be 'une simple comédie'[14] he is for a moment terrified to the point of being struck dumb. Having *just finished* the episode of the sheet which Marcelle hangs out of the window, described as 'le grand fantôme qui faisait rage dans la nuit',[15] the writer is led to associate the two images, mentioning that both are imagined or remembered as being 'on the left'. The rememoration takes place *after* the writing, therefore, and involves a re-reading of the text. There was thus 'une parfaite superimposition d'images liées à des bouleversements analogues'.[16] The two images which are superimposed on one another connect an analogous sense of terror, but whereas one is described as 'dépourvue de toute portée sexuelle',[17] the other, textual, image is 'parfaitement obscène'.[18] The relation between the two images is moreover described as a 'substitution sans aucune conscience'.[19] The textual, obscene image is a substitute, displaced but superimposable, for the childhood image which inspired only terror. One can make two remarks about this substitution. The first is that it operates spatially, or structurally, both images appearing 'on the left' and the writer emphasizing as much. Secondly, the manifestation of the traumatic childhood image is sexualized, given an obscene quality—it emerges into consciousness through its being given an obscene quality. Whereas the first point would seem to underline a parallel with the operations of the Freudian dreamwork, the second *reverses* the Freudian etiology of the symptom and the process of subli-

[13] Ibid., 74.
[14] Ibid.
[15] Ibid.
[16] Ibid.
[17] Ibid.
[18] Ibid.
[19] Ibid.

/ cure

mation: here the manifest symptom is given a sexual quality while the original trauma lacks it. In this light one would also want to note the substitution of the 'chaste' young girls and the mother of the childhood visit to the castle by the far from chaste Simone, and the description of the 'mother' as the object of an obscene aggression in the *récit*. One might explain this by proposing that the original trauma in fact had a hidden sexual content which is *released* in the manifest textual account; the effect of a repressive upbringing (which the writer mentions later) being to silence the sexual desires of the boy, a repression which is later overcome by the writing of the obscene text. I want to leave this possibility open for the moment and continue with the itinerary of the reminiscences.

2. The scene in the church in Seville has already been 'imagined', particularly the excision of the priest's eye. The writer, already conscious of the previously noted coincidence between imaginary text and the experience of his 'life', 'amuses himself' in introducing the account of Granero's death, which he had witnessed. Strangely, he has no consciousness of any link between the two, but in writing the account of the bullfight ('au moment même où j'arrivai à cette scène de la mort')[20] he is stupefied by the link between them. He reads this link as the transposition on to another person, in an imaginary scene, of the image of Granero's enucleation, 'une image qui avait sans doute gardé une vie très profonde'.[21] The text is read as the result of a direct transposition of an image which remained at a deep level of the mind on to the imaginary scene: 'Si j'avais inventé qu'on arrachait l'œil au prêtre mort, c'est parce que j'avais vu une corne de taureau arracher l'œil d'un matador'.[22] This re-reading of the text operates according to a curious temporality: the accidental encounter of an image in an American magazine leads to the realisation of a link between the imaginary text and the life of the writer. The

[20] Ibid.
[21] Ibid.
[22] Ibid., 75.

writer then amuses himself in introducing another event from his life, the writing of which reveals a further imaginary scene to have been unconsciously determined by the writer's life. 'Life', or experience, and imaginary text are thus inextricably entwined, and the process of writing of the *récit* is revealed to involve a process of analytic re-reading of itself. Again, the writer emphasizes that the transformation of the traumatic, unconscious images from 'life' emerge 'sous une forme méconnaissable'[23] due only to the obscene quality of the imaginary text, 'des que je m'étais laissé aller à rêver obscêne'.[24] Obscenity or eroticization functions as a way of overcoming the repression or blockage of the traumatic images, a way of releasing them into consciousness or memory.

3. The third incidence of revelation of the unconscious determination of the text involves a re-writing of the text in line with a series of unconscious associations, and introduces the figure of the analyst as an agent in the process of the text's composition. Hardly having finished the description of the corrida, and advised of its link to the final scene of the priest's enucleation, the writer goes to see 'un de mes amis qui est médecin'.[25] The doctor tells him that, while he had imagined the bulls' testicles to be red, and the first version of the text had described them thus, they are in fact of similar appearance to the 'globe oculaire'[26]—whitish and ovoid. Having previously imagined the bull's testicles to be independent of the chain of associations already established between eggs and eyes, the writer is thus led to suppose a level of his psyche in which these elementary images coincide. The conscious knowledge of the independence of testicles from the image cycle is thus supposed to be erroneous, a misrecognition or deformation of their actual quality, and the 'complete' cycle (as Barthes proposes it) of the *récit* is thus established—eggs, testicles, eyes—

[23] Ibid.
[24] Ibid.
[25] Ibid.
[26] Ibid.

all whitish and ovoid. This recognition presumably leads to the rewriting of the text with the actual form of the bull's testicles substituted for their original, erroneous description. In this case the text is brought in line with the revealed unconscious series of associations and transformed, owing to the introduction of a clinical voice (which one might suppose to be that of Adrien Borel); the writing of the text is again generated this time in a more explicit sense, by its analytic interpretation. However, it is not Borel, the psychoanalyst, who abstracts from this factor a metapsychological account of the functioning of the unconscious, but the writer, suggesting that the latter is himself reading the process of analysis according to his own interpretation, and not directly adopting a Freudian or other theory to account for it. He writes:

> Cette fois je risquais d'expliquer des rapports aussi extraordinaires en supposant une région profonde de mon esprit où coincidaient des images élémentaires, *toutes obscènes*, c'est à dire les plus scandaleuses, celle précisément sur lesquelles glisse indéfiniment la conscience, incapable de les supporter sans éclat, sans aberration.[27]

The theory of his own unconscious which the writer puts forward here is marked by the following qualities: a deep level of the mind which is the repository for a series of 'elementary' or 'base' images which are *all obscene*. Obscenity, in this case, is supposed as fundamental, and the text the more or less direct expression of it. The text would thus be an overcoming of the repression which would normally silence that obscenity, over which consciousness would usually 'slide'—the text takes on a transgressive status, and transgression is thereby linked to cure as the lifting or overcoming of a repression or blockage which caused the patient to act in a certain way. Conscience, however, can only support these images through a sense of rupture and aberration. The aberrant and explosive character of *Histoire de l'œil* is thus suggested to derive from the fact that consciousness, and thus language, discourse, is ruptured

[27] Ibid.

and made to 'slide indefinitely' in relation to the obscene elemental images which lie at a profound level of the mind. According to this itinerary of reminiscences the text of *Histoire de l'œil* is proposed as being on the one hand inextricably interlinked with a reading of it in relation to the author's life and unconscious, and on the other a successful overcoming of repression or blockage which silences a fundamental obscenity. At the same time, obscenity is proposed as part of a process of revelation, part of the overcoming of repression which enables the expression of latent or dormant trauma. Obscenity is at the same time the deep level of the mind which consciousness represses and cannot stand without aberration, and a process which overcomes this repression. Obscenity overcomes or exceeds its own repression, and this functions through a process of association, of 'copulation' (as proposed in *L'anus solaire*).

The writer does not stop here, however. The precision of the point of the 'écart sexuel'[28] leads to a further reading of the text, this time in relation to determinations of a specifically Oedipal character. The itinerary of traumatic or obscene, determinative images leads therefore from the reminiscence inspired by the photograph in the American magazine to the father and mother and to traumatic memories of them, an itinerary which follows the direction of Freudian analysis in its identification of the family drama as fundamental. The father is described as syphilitic, blind and suffering from terrible pain. Unable to move out of his chair, he is obliged to urinate into a 'small receptacle'. The determinative image is described thus:

> Mais le plus étrange était certainement sa façon de regarder en pissant. Comme il ne voyait rien sa prunelle se dirigeait très souvent en haut dans le vide, sous la paupière, et cela arrivait en particulier dans les moments où il pissait. Il avait d'ailleurs de très grands yeux toujours très ouverts dans un visage taillé en bec d'aigle et ces grands yeux étaient donc presque entière-

[28] Ibid.

ment blancs quand il pissait, avec une expression tout à fait abrutissante d'abandon et d'égarement dans un monde que lui seul pouvait voir et qui lui donnait un vague rire sardonique et absent (j'aurais bien voulu ici tout rappeler à la fois, par exemple le caractère erratique du rire isolé d'un aveugle, etc., etc).[29]

The writer explains that the image of his father's eyes determines the association between eyes and eggs (both being, in this case, white) which is mentioned earlier as being already constituted ('déjà ancienne'),[30] and the regular appearance of urine whenever eyes or eggs are mentioned in the text. The 'sphère métaphorique' which Barthes identifies—eyes—eggs—testicles—is thus proposed as being posterior to and less fundamental than the association of eyes, eggs and urine, that is, with the association justified by a common whiteness. The incidence of liquidity in the text is further revealed to be determined by the image of the father's eyes while urinating, rather than by the liquidity associated with the eye, its tears, for example. This suggests a conflict between a structural(ist) reading of the text as purely permutational and combinatorial and the writer's interpretation of his own text in terms of this childhood memory. The writer's interpretation *fixes* the determination of the text according to the more or less Oedipal drama of his relations to his father and to his mother.

The Oedipal character of this series of childhood memories is reinforced by the description of the following scene. The father, having gone insane, is visited by a doctor. While the doctor is conversing with the writer's mother, the father shouts out: 'Dis donc, Docteur, quand tu auras fini de piner ma femme!' The writer adds:

Pour moi, cette phrase qui a détruit en un clin d'œil les effets démoralisants d'une éducation sévère a laissé après elle une sorte d'obligation constante, inconsciemment subie jusqu'ici et non voulue: la nécessité de trouver continuellement son équiva-

---

[29] Ibid., 76.
[30] Ibid., 75.

lent dans toutes les situations où je me trouve et c'est ce qui explique en grande partie *Histoire de l'œil*.[31]

Histoire de l'œil would thus be explained by the desire to find an equivalent for the father's obscene sentence. As a transgressive text it would be that equivalent, a transgressive, instantaneous overcoming of repression experienced 'unconsciously' *up until now*, that is, up until the writing and analysis of the text reveals the unconscious motive. The father's sentence would transgress, rupture the repression of the idea of the mother as sexual object. Susan Suleiman writes, of this scene: 'what the father suddenly reveals (or recalls) to the son is that the mother's body is sexual. The knowledge that a "strict upbringing" has always tried to repress is that the mother's body is *also* that of a woman.'[32] In other words, the father reveals the truth of sexual difference. As equivalent to the father's phrase, *Histoire de l'œil* would be a transgressive revelation of female sexuality, a fascinated stare at (and from) the female sex.

The account of the father's eyes had been introduced by a countering of the usual Oedipal structure: 'à l'inverse de la plupart des bébés mâles qui sont amoureux de leur mère, je fus, moi, amoureux de ce père.'[33] This is overturned, however, when at the age of fourteen the writer conceives a profound and unconscious hatred for his father and opposes him in every way. There then follows the Oedipal drama of the father's obscene interruption. Not Oedipal/Oedipal. Subsequent to this, however, the writer considers the possible identification of his mother with Marcelle:

> Il m'est impossible de dire positivement que Marcelle est au fond la même chose que ma mère. Une telle affirmation serait en effet sinon fausse, du moins exagérée. Ainsi Marcelle est aussi une jeune fille de quatorze ans qui se trouva en face de moi pendant un quart d'heure, à Paris, au café des Deux Magots.

[31] Ibid., 77.
[32] Suleiman, 'Transgression and the avant-garde', 85.
[33] *O.C.* I, 75.

> Toutefois je rapporterai encore des souvenirs destinés à accrocher quelques épisodes à des faits caracterisés.[34]

'Marcelle *is not* my mother. This would be false, or at least *exaggerated*. Marcelle is *also* ... In any case ...' The initial denial—this is not my mother—is gradually broken down: Marcelle is an exaggerated version of my mother, Marcelle is also my mother, in any case I am going to tell you the following story which links my mother to Marcelle. The story follows: the mother, herself become insane, tries to hang herself in an attic, and then to drown herself, both of these images determining the figure of Marcelle and what happens to her: the latter particularly since, being pulled out of a stream, his mother is described as 'mouillé jusqu'à la ceinture, la jupe pissant l'eau de la rivière'. The narrative of the mother's madness and subsequent attempts at suicide follows point for point the narrative of Marcelle's insanity and subsequent death. The italicization of the text emphasizes the proximity of the two narratives and qualifies the initial restraint. We would be justified in reading this story as an expression of an aggression against the mother, and confirmed in this by the writer's *irritation* with his mother's irrationality and his violence against her:

> D'autre part, j'en arrivai à la frapper et à lui tordre violemment les poignets pour essayer de la faire raisonner juste.[35]

The following narrative thus imposes itself: love for the father (not the mother)—hatred of the father—revelation of the mother's body as sexual—love for the mother (as Marcelle)—aggression against the mother. The slippages and associations of the text cut across and subvert the broad lines of the Oedipal schema—love for the mother/hatred of the father. Like the bull's horn and the eye, or like Giacometti's *Suspended Ball* as described by Rosalind Krauss,[36] the *play* of the text dislo-

---

[34] Ibid., 77.
[35] Ibid.
[36] Krauss, *The Optical Unconscious*, 166.

cates any fixed structure and necessitates a restructuring as it moves forward. 'Coïncidences' does not, therefore, propose a hierarchy of origins, or levels of unconscious determination, with the Oedipal structure as primary, but a structural play, in which the absent 'centre' of this structure is successively displaced.[37] Any reading which would attempt to 'fix' the origin of *Histoire de l'œil* would therefore be a *restricted* reading, caught up within the text's structural play.

In this sense I want to suggest that with 'Coïncidences' Bataille is undoing ('déjouant') the code of psychoanalysis, while at the same time being caught in his own undoing, as he 'fixes' the structural play of the text in relation to his own childhood family drama. Psychoanalysis, as Deleuze and Guattari have suggested, is a powerful form of *territorializ-ation*, of fixation of the *play* of relations.[38] It fixes relations according to a certain ideological map in which the family functions as terroristic enclosure. It is a system of knowledge, a code, which organizes social space, the psyche and the body into certain zones, and imposes on the matter of experience a kind of narrative, one whose aim, ultimately, is the cure. I have already suggested how, through the use of certain figures, the discourses of psychiatry and of psychoanalysis, the one a powerful scientific and institutional presence, the other an emerging scientific *and* aesthetic context in the France of the 1920s, provide codes which the text implicitly refers to and transgresses. The psychoanalytic encounter, as we have seen, is enmeshed or included in the process of composition of the *récit*, and proposes it as part of a process of cure.

The effect of this staging of the process of writing as cure ties the text to the psychoanalytic context of its time, and to its history. The *récit* of *Histoire de l'œil* thus becomes read-able against that code and that context. 'Coïncidences' provides a framing discourse about this process, a strategy which will be adopted in all of Bataille's other fictional productions (*Le*

---

[37] Cf. Derrida, 'La structure, le signe et le jeu'.
[38] Deleuze and Guattari, *Mille plateaux*, 21–2.

bleu du ciel, Madame Edwarda, L'Abbé C) which frame their narratives with metadiscursive considerations on the nature and significance of the *récit*. We read *Histoire de l'œil* then as the staging of a psychoanalytic cure and a discourse upon it. If psychoanalysis is an interpretive, analytic strategy, in other words a process of reading, then what we read in *Histoire de l'œil* is itself a reading. The strategy of staging or framing thus functions to inscribe the psychoanalytic code in a larger textual space in which this code is rendered manifest.

The temporality of this structure is that of the *après-coup*, Freud's *nachträglichkeit*; the psyche undergoes an experience which in Bergsonian terms remains at a material level. The experience has not been experienced, but remains a solid, immobile block left in the memory. Analysis attempts the representation of the material image *after the event*, through speech or writing, a structure of *afterwardsness* which is manifest in *Histoire de l'œil*. The *récit* enables the representation, or the experiencing after the event, of those traumatic, unrepresentable images which have up until then blocked and immobilized the psyche. 'Coïncidences' thus gives an explanatory account of the ways in which the image chain of the *récit* has enabled the representation, by the writer to himself, of certain key images deriving from his childhood. But this representation is indirect—the *récit* does not itself directly represent the traumatic images; the writer, interpreting the *récit after the event*, tells us that the text has instructed him how the images of the *récit* are linked to the traumatic images of his childhood. Representation is not possible through direct expression, but only through the doubling of the *récit* by an interpretive account by the writer of the text he has already written. *Histoire de l'œil*—*récit* plus metatextual commentary—stages the process of the indirect, displaced, representation of trauma.

The cure functions, therefore, through the copulative movement of the *récit*, in which an intense and vertiginous mobility along a chain of images enables, through its very mobility and the speed of its operation, the cathecting—the representation—of traumatic images locked in the memory as indigestible blocks. Liberated from obsession and fixation, the opposite danger is thus courted; a too excessive mobility of psychic functioning. This risk has been valorized positively as that of the writer or artist, with Deleuze and Guattari's *schizanalysis*,[39] or with Kristeva's account of Bataille with reference not to the myth of Oedipus, the family drama or enclosed sexuality, but to the myth of Orestes, a psychotic traversal of boundaries.[40] What makes a writer like Bataille fascinating, however, is not only the vertiginous, potentially psychotic play of the image-chain, the unrestrained movement of the *copula*, but also the fact that this movement is staged, strategically framed and set against the codes of the *récit*, of narrative, and the various literary genres which support the text. Using an image from *Le bleu du ciel*, Troppman and Dirty's *slide* into the void, as they make love in the graveyard overlooking the city of Trieste, represents the risk and the *jouissance* of a limitless descent into the maelstrom, but the narrator's grabbing hold of a tuft of grass to check their fall figures the fragile hold on what Borel and Robin call 'reality', what Deleuze and Guattari call 'territory'.

---

[39] Cf. Gilles Deleuze and Félix Guattari, *Anti-Œdipe* (Paris: Minuit, 1972).

[40] Julia Kristeva, 'Bataille, l'expérience et la pratique', in Sollers (ed.), *Bataille*. Kristeva suggests that the Orestian writer already 'knows' the Oedipal complex, has worked through it: '... l'opération souveraine consiste à traverser l'Œdipe en représentant l'Œdipe et ce qui l'excède. Mais si l'Œdipe est la constitution du sujet unaire comme sujet connaissant, l'opération souveraine consiste à traverser l'Œdipe par un Œdipe surmontée par Oreste. Dans ses écrits de fiction, par le maintien du thème, de la lucidité et de la "méditation" comme moyens de représenter les énergies libres pré-œdipiennes, Bataille confronte Oreste à Œdipe et les met en abîme réciproquement. La traversée de l'œdipe n'est pas sa levée, mais sa connaissance' (285). In this light, cf. Louis Trente, in *Le Petit*: 'Mon père m'ayant conçu aveugle (aveugle absolument) je ne puis m'arracher les yeux comme Œdipe. J'ai comme Œdipe deviné l'énigme: personne n'a deviné plus long que moi' *O.C.* II, 60.

25. / writing blind / reading blind

What does it mean to read a text which enacts enucleation and whose guiding structural principle is an incision and castration of the eye from the head? To read, that is, to see. But reading is not only or not all a seeing; it is also a hearing, and, in a certain sense, a touching. We read the text, we read a written representation of a violence done to vision, to representation, to what makes that reading possible. Does this give an image of the text as a sublimation, as a recovery of vision after its mutilation, the text representing, restoring to visuality and to ideality a corporeal materiality of violent nature? Is Bataille, after all, on the side of sight, inoculating and making safe a disturbing threat to speculative thought from structural play, obscene excess, textual transgression? Thinking of the image of the *œil tranché* from *Un chien andalou*, the violence done to vision executed here is presented in a form which is 'beautiful', a gift to the eye in luminescent monochrome.[1] Perhaps the *image* is always a recovery of sight, a sublation of the visual: Boiffard's photography, Masson's lithographs, Bellmer's etchings would all reassert the primacy of the visual within the lines of their fixed figure. The *informe*, however, would be a discursive operation, a move in the play of writing.[2] This comes down to proposing writing, and reading, as a resistance to the recovery or sublation of sight. In their *play*, that is, in the forward movement of their structuring/destructuring, they would operate from a point of blindness, a position of risk as if at the edge of an abyss.[3] Writing would constantly over-

[1] Of course one would have to revise this critique in the light of the fact that *Un chien andalou* is a film, and the image, presented here as a still, is outlined by a *cut*, the filmic cut, which, though invisible, articulates form around its rupture.

[2] Thus the redundancy, ultimately, of mounting an exhibition of 'informal' art.

[3] My text, here and throughout, parasitically shadows Derrida's *Mémoires d'aveugle* (Paris: Louvre, 1990), which in turn proposes itself as an 'insidious' commentary on *Histoire de l'œil*. Derrida's commentary on the drawing 'of' the blind as an anticipatory (*ante-caput*: before the head) movement of an

flow into the intersticial *cut* of its differential pulse, the space open before it as it moves forward. The structuring of *Histoire de l'œil*, not only the permutative displacement of its objects but also the haunting of itself by itself that I have proposed, foregrounds this movement of anticipation and risk. The writing hand, the reading eye, would 'feel' their way forward like the hand of a blind man. Writing, or 'literature' would be that mode of play which maintained itself at this limit and refused to enclose itself or to resolve itself within the fixed contours of *an* image.[4] *Histoire de l'œil*, then: the story of the eye as it moves its way forward like a blind hand through the text, cutting through and across, in its play, the images of the eye of vision.

---

imaginary hand with an eye open at the fingertips (a supremely 'Bataillean' image) offers itself also as a commentary on *writing*. Writing, in this case, would correspond to the tracing of a line, not a figure or a form, or an *out*line. As such Masson's drawing, Bellmer's etchings 'au *burin*' (see p. 99) *would* assert themselves, as Leiris's description of Masson's art suggests, as an *exploration of the line*, the tracing of a mobile, inventive and 'blind' line. Perhaps a line of flight?

[4] Hollier, *La prise de la Concorde*, 54: 'Peut-être l'œuvre de Bataille trouve-t-elle sa plus grande force dans ce refus de la tentation de la forme ...'

Bibliography

Georges Bataille, *Œuvres complètes*, vols I–XII (Paris: Gallimard, 1970–95).
—— *Story of the Eye*, trans. Joachim Neugroschel (Harmondsworth: Penguin, 1979).
—— *A Tale of Satisfied Desire*, trans. Austryn Wainhouse (Paris: Olympia Press, 1958).

Select bibliography of critical works on Histoire de l'œil *and related texts*

Barthes, Roland, 'La métaphore de l'œil', in *Essais critiques* (Paris: Seuil, 1966).
—— 'Les sorties du texte', in P. Sollers (ed.), *Bataille* (Paris: 10/18, 1973).
Didi-Huberman, Georges, *La ressemblance informe ou le gai savoir visuel selon Georges Bataille* (Paris: Macula, 1995).
Dworkin, Andrea, *Pornography: Men Possessing Women* (New York: Perigee, 1981).
Fédida, Pierre, 'Le mouvement de l'informe', in *La part de l'œil 10: Bataille et les arts plastiques* (Bruxelles: *La part de l'œil*, 1994).
Fitch, Brian, *Monde à l'envers, texte réversible: la fiction de Georges Bataille* (Paris: Lettres Modernes, 1982).
Hollier, Denis, *La prise de la Concorde* (Paris: Gallimard, Collection 'Le Chemin', 1974), in English as *Against Architecture*, trans. by the author (Cambridge, Mass.: MIT, 1992).
—— 'Bataille's tomb: a Halloween story', *October*, 33 (Summer 1985).
Krauss, Rosalind E., *The Optical Unconscious* (Cambridge, Mass.: MIT, 1993).
—— 'Corpus Delicti', in *L'amour fou: Surrealism and Photography* (New York: Abbeville Press, 1985).
Krauss, Rosalind E. and Bois, Yves-Alain, *Formless: A User's Guide* (New York: Zone Books, 1997).
Masson, André, *Le rebelle du surréalisme: Ecrits* (Paris: Hermann, 1976).
Matossian, Chakè, 'Le rat et l'œuf (Bataille, l'*Histoire de l'œil* et le clin d'œil de Valdès Léal)' in *La part de l'œil 10: Bataille et les arts plastiques* (Bruxelles: *La part de l'œil*, 1994).

Popowski, Michael J., 'The eye of illegibility: legibility in *Histoire de l'œil*', in Leslie-Ann Boldt (ed.), *On Bataille* (Albany, N.Y.: State University of New York Press, 1995).

Sontag, Susan, 'The pornographic imagination', in *Styles of Radical Will* (New York: Delta, 1981).

Steinmetz, Jean-Luc, 'Bataille le mithriaque (sur *Histoire de l'œil*)', in *Revue des sciences humaines*, 206 (1987).

Stoekl, Allan, *Politics, Writing, Mutilation: The Cases of Bataille, Blanchot, Roussel, Leiris and Ponge* (Minneapolis, Minn.: University of Minnesota Press, 1985).

Suleiman, Susan, 'Transgression and the avant-garde: Bataille's *Histoire de l'œil*', in *Subversive Intent* (Cambridge, Mass.: Harvard University Press, 1990).

Taylor, Mark C., *Altarity* (Chicago and London: University of Chicago Press, 1987).

Warin, Francis, *Nietzsche et Bataille: la parodie à l'infini* (Paris: PUF, 1994).

Index

45 Rue Blomet, 58–9, 61–6, 67 n 13,
68–9, 68 n 14, 71, 76–8
54 Rue du Chateau, 58–61, 64
Acéphale, 77, 80, 80 n 2, 133, 140 n
25, 150
'Aimée', 128 (see also Lacan,
Jacques)
Angélique, Pierre, 36 n 8 (see also
Bataille, Georges)
Anthropology, 58
Apollinaire, Guillaume, 60, 65–70,
67 n 10, 89, 104
Aragon, Louis, 33 n 1, 59, 60, 66, 68,
75 (see also *Le con d'Irène*)
Aréthuse, 34, 44–7, 49, 52, 57, 159
Arrabal, Fernando, 40, 40 n 6
Artaud, Antonin, 58, 59, 66, 75
'Audiart', 36 n 8 (see also Wainhouse,
Austryn)
Aury, Dominique, 39 n 5 (see also
Réage, Pauline)

Babelon, Jean, 47, 57
Balzac, Honoré de, 14, 151–4, 152 n
9, 155 n 19
Bar du Chateau, 58 (see also 54 Rue
du Chateau)
Barthes, Roland, xi, xi n 5, 5–16, 5
n 3, 18, 18 n 3, 24, 26, 36 n 8,
40, 78, 94, 97, 98, 146, 151, 166,
169
Bataille, Georges (works other than
Histoire de l'oeil)
L'Abbé C., x, 173
'L'Amérique disparue', 45, 47–9,
76
L'anus solaire, 19, 34, 38, 44, 45,
52–7, 72, 75, 123, 140, 154 n
16, 168
Le bleu du ciel, ix n 1, x, 15, 35,
44, 59, 61, 84, 84 n 3, 133, 144,
150, 173, 174
Choix de lettres, 158 n 3
L'érotisme, 37, 109, 125, 133

L'expérience intérieure, 34, 38, 143
n 37
La littérature et le mal, 15, 105
Ma mère, x
Madame Edwarda, x, 35, 35, 42,
36 n 8, 128, 154, 173 (see also
Angélique, Pierre)
Manet, 111
Le mort, 35
'La notion de dépense', 34
Notre Dame de Rheims, 34, 43, 48,
50, 52, 112, 133, 159
Oeuvres complètes, 35
L'ordre de la chevalerie, 43, 44,
45, 52
La part maudite, 34
Le petit, 79, 120, 158 n 3 (see also
Trente, Louis)
'Plan d'une suite à *Histoire de
l'oeil*', 37
'Préface à Histoire de l'oeil', 30 n
9, 79, 120, 158 n 3
Sacrifices, 75
W.C., 30 n 9, 51, 59, 79, 120, 160,
159 n 3
Bataille, Martial, 158–9
Baudelaire, Charles, ix n 2, 107–11,
113, 114
Beaumarchais, Pierre Augustin
Caron de, 133, 153, 153 n 15
Bellmer, Hans, 1 n 4, 34, 35, 40, 74,
175, 176 n 3
Benjamin, Walter, 63 n 8
Bergson, Henri, 161–162, 173
Bernini, Giovanni Lorenzo, 128, 129
Beuys, Joseph, 24
Bibliothèque Nationale, 43, 44, 56,
57, 67 n 1 3, 111, 119, 123
Blanchot, Maurice, ix n 1, 77, 124
Bloch, Ivan, 69
Bloom, Harold, 17, 39, 39 n 2
Boiffard, Jacques André, 144, 175
Bois, Yve-Alain, 17, 25, 27
Boldt, Leslie Ann, 30 n 11
Bonaparte, Marie, 127 n 12, 160
Bonnel, René, 33 n 1, 60, 68

Borel, Adrien, 28 n 4, 44, 52, 130, 159, 160–1, 167, 174
Bourgeois, Louise, 24
Boyars, Marion, 36 n 8
Brakhage, Stan, 1
Braque, Georges, 65, 66, 70
Breton, André, 44, 58–60, 63–7, 67 n 11, 69, 71–4, 75, 106–7, 114, 127, 129, 140 n 25, 160–1
British Museum, 57
Buchloh, Benjamin, 23–5
Buñuel, Luís, 1 n 4, 1, 3, 60, 99, 115 n 4, 127, 129 (see also *Un chien andalou*)
Bureau des recherches surréalistes, 59, 62
Byron, George Gordon, Lord, 149, 150, 150 n 7

Carter, Angela, 40
Cézanne, Paul, 72, 74
Chapsal, Madelaine, 158 n 3
Char, René, ix n 1
Charbonnier, Georges, 61
Charcot, Jean Martin, 127, 128 n 7
Chestov, Léon, 44, 61
Un Chien Andalou, 1–3, 32, 60, 101, 129, 175, 175 n 1 (see also Buñuel, Luís, and Dalí, Salvador)
La Ciguë, ix n 1
Cinema, 1
Cohen, Margaret, 63 n 8
College de Sociologie, 46
Le con d'Irène, 60 (see also Aragon, Louis)
Corday, Charlotte, 127 n 12
Critique, ix n 1, xi n 5, 36 n 8, 39 n 5, 133
Cubism, 64–7, 70, 71, 74–5, 77, 78

D'Espezel, Pierre, 47, 57
Da Ponte, Lorenzo, 133
Da Vinci, Leonardo, 109–10
Dada, 68, 74
Dalí, Salvador, 1, 1 n 4, 3, 60, 64, 74, 99, 115 n 4, 127, 129, (see also *Un chien andalou*)
Damiens, 118 n 16
Darsat, 89

Dausse, Dr., 52, 159
De Chirico, Giorgio, 64, 74
De Mandiargues, André Pieyre, 39–40, 39 n 3
Deleuze, Gilles & Guattari, Félix, 22, 48–9, 51–2, 57, 172, 174
Dépense, 140 n 25
Derain, André, 66
Derrida, Jacques, 8, 12, 13, 32 n 16, 54, 103, 131, 175 n 3
Des Forêts, Louis-René, ix n 1
Desnos, Robert, 59, 60
Dick, Leslie, 127 n 12
Didi-Hubermann, Georges, 4, 5, 18–19, 26–30, 28 n 4, 30 n 9, 128 n 7
Dispot, Laurent, 117 n 14, 121 n 22
Documents 2, 4, 6, 17–19, 21, 26, 45, 46, 32 n 15, 47, 49, 41 n 13, 58–60, 68, 119, 140, 143 n 33, 144, 160, 155 n 19
Don Juan, 34, 35, 64, 87, 89, 92, 95, 104, 116, 132–3, 139, 143, 148–53
Don Miguel de Manara, 149, 150, 150 n 7
Dostoyevsky, Fyodor, 61, 62
Dubuffet, Jean, 66
Ducasse, Isidore, 67–69
Duchamp, Marcel, 20, 68–70, 103, 104 n 1, 116, 121
Duhamel, Marcel, 59
Dühren, Eugen, 69
Dumézil, Georges, 46, 48
Dupin, Jacques, 77
Duras, Marguerite, ix n 1

Ecole des Chartes, 58
Eisenstein, Sergei, 4–6, 5 n 3, 15
Eluard, Paul, 127, 129
L'évolution psychiatrique, 160

Fautrier, Jean, ix n 1
Fédida, Pierre, 18, 18 n 4, 28 n 4, 30 n 9
Ffrench, Patrick, x n 3, 63 n 8
Florange, Charles, 49
Foucault, Michel, ix n 1, 35, 37, 103, 119

France, 34, 48, 50, 84, 84 n 3, 87, 132
Freud, Sigmund 13, 28 n 4, 30 n 9, 31 n 12, 38, 72, 80, 80 n 4, 107, 131, 143, 160–4, 163 n 11, 167–8, 173
Futurism, 67

Houdebine, Jean-Louis, 60 n 1
Hugo, Victor, 155 n 19

Informe, 17, 54, 75, 78, 105, 113, 144
Irigaray, Luce, 30 n 10

Gender, 136
Genet, Jean, ix n 2, 127 n 8
Genre, 85
Gérard, Francis (Gérard Rosenthal), 59
German Romanticism, 81 n 4
Giacometti, Alberto, 20, 50, 116, 121, 137 n 21, 171
Girodias, Maurice, 36 n 8
Godard, Jean-Luc, 41, 42
Goncourt, Edmond & Jules de, 118 n 16
Gorky, Arshile, 72
Gothic Novel, 85, 115, 124, 128–9, 132, 154, 156, 163
Goux, Jean-Joseph, 80 n 2
Goya, Francisco de, 111, 146 n 52
Grandville, J.J., 138, 155 n 19
Granero, Manuelo, xi, x, 9, 30, 59, 84, 90, 95–7, 102 n 3, 135–9, 141, 146–8, 150, 165
Griaule, Marcel, 58
Gris, Juan, 64, 66, 71, 74
Guibert, Hervé, 40, 40 n 8
Guillotin, Joseph-Ignace, 117
Guyotat, Pierre, 41

Jacob, Max, 60, 65–9
Jakobson, Roman, 8, 9
Janet, Pierre, 161
Jarry, Alfred, 67–9, 121
'Le jésuve' , 154 n 16
Jouhandeau, Marcel, 66
Jouissance, 15–17, 54, 76, 78, 98, 128, 174

Kafka, Franz, 121–2, 124
Kahane, Maurice, 36 n 8
Kahnweiler, Daniel, 76
Kaplan, Jo-Ann, 1 n 4
Klossowski, Pierre, ix n 1, 39, 39 n 4, 40, 80 n 2
Knight, Diana, 24 n 16
Krafft-Ebing, Richard, Baron von, 128
Krauss, Rosalind E., 17–27, 18 n 3, 50, 60 n 2, 77, 78, 104 n 1, 119 n 20, 137 n 21, 171
Kristeva, Julia, 8, 12, 21–25, 24 n 15, 25 n 18, 174, 174 n 49

Hatoum, Mona, 24 n 15
Haunting, 155–6
Hegel, Georg Wilhelm Friedrich, 18, 27
Heidegger, Martin, 13
Heilbrun, Georges, 69
Heine, Maurice, 69, 75 n 38
Hemingway, Ernest, 58, 133, 133 n 1
Heraclitus, 56, 63
Holbein, Hans (The Younger), 24 n 17
Hollier, Denis, ix n 1, 44, 48, 49, 50, 107, 112, 119 n 20, 158 n 3, 176 n 4

Lacan, Jacques, 24, 24 n 17, 80 n 2, 127 n 8, 128–30
Lanchner, Carolyn, 75 n 37
Lautréamont, Comte de, 75, 115, 122 n 24, 123, 154 (see also Ducasse, Isidore)
Lavaud, Jacques, 57–9
Leiris, Michel, ix n 1, 36 n 8, 44, 45, 45 n 7, 57–64, 66, 67 n 13, 68 n 14, 70–4, 77, 99, 114 n 34, 122–4, 133–4, 138, 160, 176 n 3
Lely, Gilbert, 69
Limbour, Georges, 58, 61, 63, 66, 75
Lombroso, Cesare, 128

Lord Auch, x, 34–7, 79, 80–3, 98, 157, 158 n 3, 159 n 3

Orphism, 67

Magritte, René, 64, 74, 109 n 20
Mallarmé, Stéphane, 65–6, 73, 75, 121–2
Malraux, André, ix n 1, 84 n 3, 111–12, 133
Man Ray, 1 n 4
Manet, Edouard, 38, 108, 110–14, 146 n 52
Marx, Karl, 80 n 2
Masson, André, 33 n 1, 34, 40, 58, 60–4, 66–9, 67 n 13, 68 n 14, 69 n 16, 71–8, 80 n 2, 122, 133–4, 150, 175, 176 n 3
Mathews, Harry, ix n 1
Matossian, Chakè, 149 n 5, 149 n 6, 155 n 19
Maturin, Charles, 153 n 15
Mérimée, Prosper, 133
Métraux, Alfred, 58
Meunier, Victorine, 113–14
Michelet, Jules, 15
Mimesis, 84
Minotaure, 127 n 8, 140 n 25
Mirbeau, Octave, 124, 129
Miró, Joan, 40, 58, 63, 64, 66–7, 71, 74, 77, 78, 120
Mishima, Yukio, 40
Molière, 133, 150
Montherlant, Henri de, 84 n 3, 133
Morris, Frances, 24 n 15
Mozart, Wolfgang Amadeus, 133, 150
Murillo, Bartolomé Esteban, 149, 149 n 6
Musée de l'Homme, 58

Names, 100, 105
Nancy, Jean-Luc, 30 n 10, 32 n 16, 33, 35, 80 n 4
Neugroschel, Joachim, ix n 1, 36 n 8
Nietzsche, Friedrich, ix n 2, 13, 44, 61, 118 n 17, 119, 120, 143
Nöel, Bernard, 40, 41
Nord Sud, 59
Nozières, Violette, 128

Papin, Christine and Léa, 27 n 8, 128
Parody, 52, 53, 55, 93
Paulhan, Jean, 39 n 5, 67, 69, 67 n 12
Pauvert, Jean-Jacques, 35
Picasso, Pablo, 66, 70, 75, 133
Pleynet, Marcelin, 40, 40 n 9
Poe, Edgar Allan, 115, 124
Pollock, Griselda, 25 n 18
Pollock, Jackson, 72, 75
Ponge, Francis, 124
Popowski, Michael J., 30 n 11
Pornography, 80, 85
Praz, Mario, 118 n 16, 153 n 15
Prévert, Jacques 59, 60
Proust, Marcel, 86, 105, 159
Psychoanalysis 56, 93, 99, 130, 154
Punctum, 6, 146

Queneau, Raymond, 59, 160

Racine, Jean, 140 n 25
Radcliffe, Mrs., 153 n 15
Réage, Pauline, 39, 39 n 5, 40
Reboul, Jean, 15 n 2
Reverchon, Blanche, 161
Reverdy, Pierre, 59, 65
La Revolution Surréaliste, 44, 60
Rimbaud, Arthur, 51, 51 n 26, 67
Rivière, Georges-Henri, 58
Robin, Gilbert, 160, 161, 174
Rodin, Auguste, 71
Roudinesco, Elisabeth, 160 n 4
Roussel, Raymond 58, 67–70, 67 n 13, 68 n 13, 99, 104, 121–4
Rubin, William, 75 n 37
Rue Fontaine, 58, 61–3

Sade, Donatien Alphonse Francois, Marquis de, 36 n 8, 39 n 4, 40, 41, 49, 61, 67–70, 69 n 16, 75, 75 n 38, 76, 89, 106, 106 n 12, 107, 115, 115 n 4, 124, 126, 163

Saint Paul, 45
Salacrou, Armand, 66
Salpetrière, Hopital de la, 128
Sartre, Jean-Paul, ix, ix n 2
Saussure, Ferdinand de, 80 n 2
Schaeffner, André, 58
Siena, 59
Snow, Michael, 1
Société Psychanalytique de Paris,
    160
Sollers, Philippe, ix n 1, 40, 60
    n 1
Sontag, Susan, 36 n 8
Soupault, Philippe, 66, 71
Spain, 35, 84, 87, 84 n 3, 89, 90, 132,
    133, 138, 142, 143, 150
Stein, Gertrude, 58, 73
Steinmetz, Jean-Luc, 116 n 8, 121 n
    22, 152 n 9
Stoekl, Allan, 123
Suleiman, Susan, 110
Surrealism, ix, xii, 57–68, 71–4,
    78–9, 81 n 4, 104 n 1, 121–3, 127,
    129, 140 n 25
Surya, Michel, 130 n 16
Sylvester, David, 71 n 17

Tanguy, Yves, 59, 74
Tel Quel, ix n 1, 12, 14, 40, 40 n 9, 41
Thirion, André, 59
Tirso de Molina, 150
Tolstoy, Lev, Count, 44
Transgression, 53
Trente, Louis, 79, 80, 120, 158 n 3,
    159 n 3, 174 n 49 (see also
    Bataille, Georges)
Tual, Roland, 58, 60, 66, 69
Tzara, Tristan, 66, 68

Valdés Léal, Juan, 149, 149 n 6, 150
    n 6
Valéry, Paul, 111
Vertige, 95, 97, 101, 126
Villon, François, 145 n 44

Wainhouse, Austryn, ix n 1, 36 n 8
Walpole, Horace, 115, 153 n 15
Warhol, Andy, 1
Warin, Francis, 118–19, 118 n 17
Wyn Evans, Cerith, 1 n 1

Index   183